to Ted, Thank you
for your interest,

HIGHER INTELLIGENCE

How To Create A Functional Artificial Brain

HIGHER INTELLIGENCE

Peter AJ van der Made

Published by Vivid Publishing
P.O. Box 948, Fremantle
Western Australia 6959
www.vividpublishing.com.au

A cataloguing record of this book is held at the National Library of Australia.

To order further copies of this book or to contact the author, please visit www.vividpublishing.com.au/higherintelligence for further information.

Contents

About the Author

There is little or no agreement on what intelligence is. At this time neither Artificial Intelligence technology, nor the impact of that technology on human society are well understood. So called 'smart' devices are not smart, let alone intelligent. What is being promoted as 'smart' or 'intelligent' is nothing but a programmed cause-and-effect machine that repeats the same routines each time the same stimuli are applied. Artificial Intelligence, in the sense of a machine that is aware and capable of making intelligent decisions on indefinite criteria, does not yet exist. What needs to change to make intelligent machines a reality? Why has this not happened already? This book may be the first step in the direction that leads to a truly intelligent, evolving and learning machine.

Since the mid-1940's books and movies have been published about intelligent robots by such well-known authors such as Asimov and Heinlein. All over the world attempts are underway to make fiction a reality by attempting to understand the brain, and program its models on large computers. The brain is far superior to a computer in cognitive processes, and a computer is far superior to the brain in performing mathematics operations. Rather than to follow in everyone's footsteps and write programs, my passion is to build learning, cognitive machines in hardware using the brain's circuits as a model. There are good reasons to use dedicated hardware. The brain is not a sequential machine, nor does it work by digital arithmetic. Starting off with a broad experience base built over 40 years in computer science I initially used an array of RISC processors. Four years into the project I realized that processors are badly structured to emulate brain functions; computers are calculating machines, and the brain is nothing like a calculator. After this realization I started to design specialized digital circuits that simulate the analog processing methods of the brain. Over the years I have improved on the original inspiration.

Some people are fortunate enough to turn a passion into a career. I am one of them. My interest in computer technology started in the late 1960's. The press was describing these machines as 'giant brains' and I wanted to find out how they worked. As a boy I was often found at the local library or at the Evoluon in Eindhoven, not far from my home town in the Netherlands. The Evoluon was a popular science and technology exhibition that opened its doors in 1966. Unfortunately, the building closed down in 1989 due to a lack of interest. I think that 'lack of interest' can be interpreted as a 'lack of vision and renewal'. Kids will always be interested in cutting edge technology that is way beyond what they can find at the local mall or online. They want to experience it, to be able to feel it, touch it and play with it. An interactive technology exhibit needs to have things like a quantum computer, an uplink to a supercomputer, robotics, and artificial intelligence technology. Simple push buttons are not seen as 'interactive' any more – if they ever were – kids need to have access to different levels of programming and game-like interfaces. Every town should have a place like this. The Evoluon exhibit inspired me to study electronics engineering. I was fascinated by the way digital electronics works, how a bunch of simple logic gates operating on nothing more than ones and zeroes could be made to perform complex mathematics, how programs were built and compiled. My earliest programs were written in the early 70's in Fortran IV for the large Burroughs mainframe computer that we had at school. Early microcomputers based on the Intel 8008 and 8080 used Assembler, a language close to machine language. These machines used the Digital Research CP/m ™ operating system I had to write a driver in Assembler to add a new device like a hard disk drive (yes, they did exists in the 70's). I built my first home computer in 1976, a 40x40x60 blue (what else?) box with 8 inch disk drives and a 'huge' 5 Mb hard disk. Using this blue box as a development system, the first 900 by 1200 color graphics accelerator card for the IBM PC was created in 1982. The original IBM-PC was running at 4.77 MHz, and my graphics processor was pushed as high as 120 MHz. A full-custom (ASIC) chip was made of this design at the beginning of 1986. After a time teaching computer science at a university I came across a new problem that needed my attention. A solution was needed for computer viruses. Viruses had become a persistent problem and cost businesses millions of dollars a year. Scanning for virus signatures lagged behind by months and did not prevent widespread infections with new viruses. My computer immune system used an emulation technology that I patented in the US, called an Analytical Virtual Machine. This system detects malware by analysis long before a signature becomes available. The rights were sold to a leading computer security company in 2002.

This gave me the opportunity to research the brain and to learn how to build cognitive machines. What has emerged from that research is a digital emulation of cells that make up the brain, with synapses and various neuro-transmitters. Only 6500 logic gates make up a single node, including IO and synaptic memory. I call it a 'Synthetic Neuro-Anatomy' to set it apart from old fashioned neural networks. The exciting part is that it learns, just like the brain learns, from sensory input. The learning abilities of the machine thrilled me when I tested the prototype in a field programmable gate array (FPGA). This is far more efficient than using software. Just consider the millions of gates that make up a microprocessor, not to mention the memory and I/O circuits.

Even though this is new technology, it uses standard silicon manufacturing techniques but no processor or programming in the traditional sense. It learns to perform a task by being exposed to sensory stimuli that are specific for the task. The memory image that results from learning, the 'Training Model' can be saved in a library and reused in other machines. Native learning machines present a new direction in Artificial Intelligence.

New materials and methods for computer technology generally take a long time in the laboratory. We have been reading about Quantum Computing since the middle of the 70's but more than 40 years later the technology is still in its infancy. It will be a while before we get to play with a quantum home computer. Carbon nanotube transistors have similarly been trapped in the laboratory stage. By using standard silicon manufacturing techniques it is possible to bring intelligent devices to market very quickly. Early adopters could be experimenting with synthetic brains within a year.

What shape will intelligent systems take? Most likely they will be embedded in things that we don't recognize as a 'computer', such as intelligent toys that learn as you interact with them or a safety device in your car that learns how to drive and warns you, or limits dangerous driving conditions. Eventually we will learn how to build elaborate robots that are able to be taught a task, after we upload training models from libraries of simple functions. With a new technology like this, it is difficult to look over the horizon of its potential.

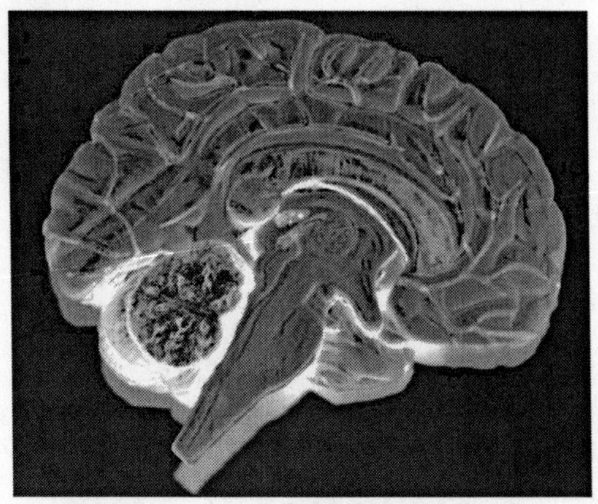

Cross-section of the human brain

Image created by the author © 2011

The origin of love, intellect and creativity is in the spiritual, the invisible regions of our mind. The mind is not the brain. The brain is the information carrier; the mind is a function of brain neuron distribution and what is stored there. 'Knowledge' is not just data, it is relational in illogical ways and it determines the brain's algorithms. The mind grows over time by what it feeds on. It is who we are, all our accumulated experiences, beliefs, joy, fears, anguish, love, scars, and knowledge. The sum of all our experiences makes up our intellect. A learning machine of sufficient complexity and structure, given the right stimuli, will eventually become an intellect.

Peter AJ van der Made

Introduction

"Every man can, if he so desires, become the sculptor of his own brain".

Santiago Ramón y Cajal

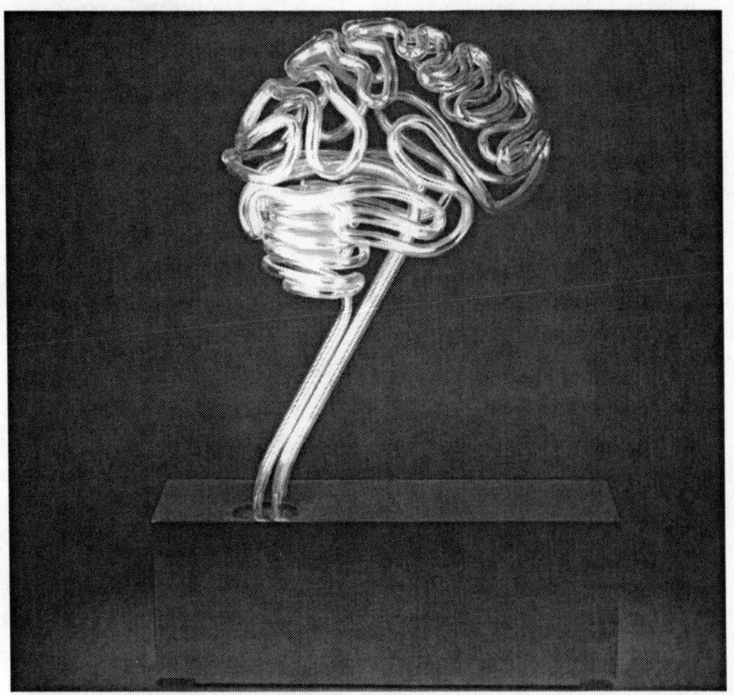

« Tree of science – 13 000 Volts » by Cedric Ragot / light sculpture 2010 credit photo : Tristan Everhard

Artificial intelligence is a branch of computer science. Even though the rest of the computer field has seen astonishing progress over the last 65 years, the field of Artificial Intelligence has advanced very little. Intelligence has proven difficult to compute, far more difficult than prominent computer scientists such as von Neumann or Alan Turing could have foreseen. It is time for a new approach to the problem, a technology that does not replace computers but augments their functionality and cognitive processing capabilities. The human brain can be described in several ways. It is a giant 3D storage facility containing highly relational information. In another way it is a self-organizing association-based recognition system. It is a creative entity that can combine thoughts in a seemingly random way to come up with new ideas. As a storage system, it can store more than 100 Terabytes of highly interconnected, relational information. It most significant feature is that it learns. The brain does not run any programs. The previously stored (learned) information determines how incoming information is processed. If effect, the stored information forms its program that determines things like the path of movement of a limb, the pronunciation of words, social behavior and everything else we do. Information is stored in a way that is unlike data in a computer memory. It is stored as a value in a chemical node in analog form and it is constantly updated. Computer data is static, while the information in the brain is fluid. The incoming sensory information stream is used to recall information from its analog memory, which in turn addresses all the other associated information in a method that is similar to a linked list. The information is refreshed when it is accessed and new information is added. The brain has a prediction mechanism that uses the incoming sensory information to guess what comes next. Because of the huge differences between the brain and digital computers, a gap exists that has proven impossible to span by programming or clever shortcuts. Creating brain functions in hardware, rather than in software, adds a new dimension which is impossible to create in computer software; the synthetic brain is freed from the constrains imposed by a computer's sequential processing cycle, limited I/O space and narrow data bus. The brain's data bus is millions of bits wide. Learning is a prime function of the brain that has been absent in most brain emulations up to now. It is the process by which a brain achieves a purpose. Learning is the way by which the brain builds the mind. The mind has a multi-dimensional pyramidal organization, with simple functions at its base and functions of increasing complexity layered on top.

The human mind is the only intelligent entity in the known universe. It is contained in the brain 'hardware', which is a complex organization of an estimated 86 billion neural cell nodes[1] and glial cells. Each cell has thousands of synapses. Synapses are both connection points and memory locations. There are many different classifications of cells but the two main groups are neural cells and glial cells. Neural cells are similar to the other cells in the body, but they communicate with one another through massive parallel connections that carry electrical pulse streams. Such cells are found everywhere in the body, but especially in the brain. In addition to these cells there are in excess of 100 trillion synapses – tiny electro-chemical memory 'knobs' that cover the cells. Most neurologists agree that synapses are where our memory resides. Synapses are found in great numbers attached to every cell of the brain, so memory is distributed throughout the brain. It is these humble synapses that contain our mind.

---

[1] S. Herculano-Houze. 2009. The Human Brain in Numbers.

It makes sense that the design of an intelligent machine, which is a cognitive machine that can learn, think and is aware of its environment, should be inspired by the brain. This has not been the case over the many years of Artificial Intelligence research. Engineers have attempted to reinvent intelligence and tried to confine it inside a rigid computer framework. Considering the long history of failures it does not look like the recreation of intelligence will produce any results. Copying the processes that occur in brain may be the only way that intelligence can evolve, once we agree on what intelligence is. The meaning of the word remains contentious. It has been abused by marketing simple gadgets, cause-and-effect machines, as 'intelligent'. This has led to an understanding of "Intelligence" that is different from human intelligence. Cause and effect machines, whereby the same inputs lead to the same action cannot be called intelligent. To avoid confusion of 'new' A.I. and the old failures, the latest incarnation of A.I. had been dubbed "Artificial General Intelligence" or AGI. Intended to avoid the mistakes of the past they still continue along the same old trail of attempting to reinvent intelligence and to create it programmatically in a digital computer environment. Brain emulation at the neuron level requires huge amounts of processor power, about a hundred million times the power of a modern desktop PC. Supercomputers are large machines that calculate at incredible speeds and have huge amounts of storage. Fujitsu's K-computer attained 10.51 PetaFlops (10,510,000,000,000,000 calculations per second). It consists of 88,128 processors. The race is on for the first 100 petaflop computer. A basic human brain, consuming no more than 20 watts of power, outperforms any supercomputer in tasks that involve intelligence or recognition. The brain does not have any processors, nor does it run programs. Its highly parallel structure is one of the many strengths of the brain. Say at any given millisecond in time, 15% of the brain's circuits are 'active'. That represents 15 billion neural cells that are processing information stored in some 15 trillion memory locations simultaneously. That is a throughput equivalent to 15 Petabytes (Peta = 1000 Terra or 1,000,000 Gigabytes) of relational data per second. But its parallel processing speed is not the only strength of the brain. The comparison of logic gates to neurons was based on the early misconception that a neuron is a switch. We know now that the processing algorithm of a neuron is so complex that it cannot be compared to a to a single floating point operation (FLOP) let alone a switch. Therefore, the comparison of an array of 86 billion neurons, like the human brain, to a 10 Petaflop supercomputer is meaningless. Software determines the functions that are performed by such supercomputers.

Fujitsu's K-computer comprises 864 racks. Image: © Fujitsu

Neurons are signal processors in which the content of the synapses, the input pattern over time (temporal-spatial input pattern) and its history change the pulse properties of the output signal. A single neuron can have up to 200,000 synaptic storage locations. A neuron has an autonomous learning ability, and simultaneously integrates sequences across thousands of channels and over time to determine if the temporal pattern at its inputs is similar to a pattern that was previously learned.
A neuron, its supporting glial cells and its synapses is an analog computing node with a response time of 1-2 milliseconds. Its input to output behavior is well documented. Combinations of these analog computing nodes perform a function such as the recognition of a line segment or a single audio frequency. A functional block consists of about 10,000 neurons and is referred to as a 'Column'. Accurate software simulation of the properties of a neuron and its synapses takes much computer time. Limited neuron models are inadequate to perform learning and cognitive functios.

By way of comparison, the logic gates in a computer are simple switches, they switch between a one and a zero, depending on the combination of ones and zeros at the inputs. A logic gate switches from logic 1 to logic 0 when both its inputs are logic one. All the complex functions in a computer are built from different combinations of these simple components. Two logic gates can be combined to create a simple 2-state memory. Three of these logic gates make a 1-bit adding circuit. A microprocessor consists of many millions of such gates (see appendix 2for more details).

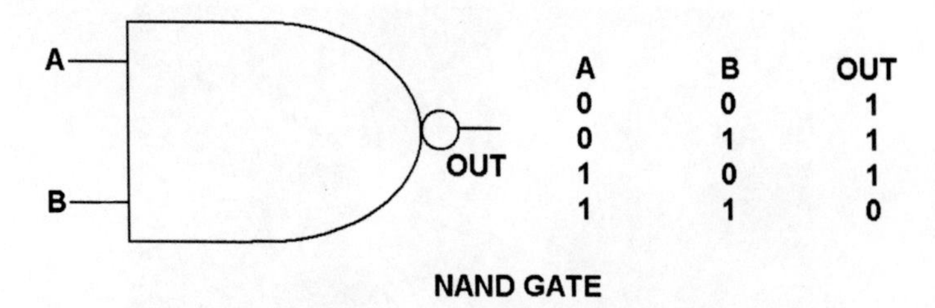

A	B	OUT
0	0	1
0	1	1
1	0	1
1	1	0

A symbol for a computer logic gate and its behavior table.

To think of a neuron as a switch is ludicrous. Neurons are nothing like computer gates. They do not switch through a combination of ones and zeros on their inputs, even though 'old' artificial neural networks (ANN) worked in this way. Ants communicate with one another through a pattern of movements. Neurons behave somewhat like ants, that communicate with one another through movements. Neurons communicate with one another through pulse streams. Neurons pass messages in the shape of timed electrical impulses in which the frequency, the pulse intensity and interval timing are all important factors. Even though the brain is complex, the individual messages between neurons are relatively simple. An impulse causes a chemical change in a connection point, a synapse. The synapse contains a chemical memory.

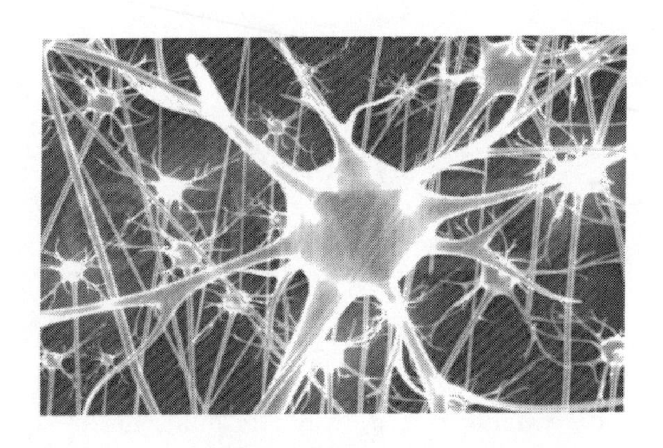

Neurons with dendrites where thousands
of synapses attach
Image © royaltyfreeimages.com

The average number of synapses per cell is about 7000. The output from a cell is an electrical pulse, or a sequence of pulses, which represents the temporal (sequence over time) memory stored in the cell. The memory is recalled by the sequence of pulses at its input. The output connects to other cells, causing a similar response in them.

The brain starts learning from the moment it is formed and continues to learn over a lifetime. At birth, the brain contains innate 'pre-programmed' knowledge, placed there by DNA. A machine that is structured like the brain and has the same mechanisms as the brain learns by being exposed to the environment through its sensory devices. It also has the same feedback paths that are part of the brain's learning mechanisms. Such an intelligent machine will need to learn before it can do anything. Learning ability relies on being synchronized to the real world. It can only add to its store of experiences and real-world objects when it can interpret the world at its own pace. How well the brain's structure is formed, which is defined by

DNA and several other factors such as oxygen levels at birth, determines how well it learns but not the learning method. The next hurdle on the path to machine intelligence is to get the structure right to allow intelligence to emerge through learning.

The advantages of building intelligent machines are evident; such machines are able to take over dangerous, prolonged or repetitious tasks that now only humans can do, such as dangerous exploration work, working in underground mines, disabling explosive devices, fire search and rescues, space exploration and search for, and destroy landmines. We can build intelligent toys that keep an eye on the kids, and machines that sample their environment to change their behavior. We can build sniffing devices that are better than sniffer dogs to detect cancer or drugs, and many other things that we have not even thought about. We can build such machines today. The idea that is presented here is revolutionary; a machine that learns and autonomously evolves intelligence. The daunting bit is to let go of the inbred method of program control. Intelligence evolves by building upon acquired knowledge in a never-ending cycle of learning and applying learned knowledge. Knowledge that has been acquired and built upon has a depth and complexity that programmed knowledge cannot achieve. Programmed knowledge is static, like the information stored in a book. Any system that does not depend on learning to increase its knowledge is a cause and effect controller.

Contrary to popular belief, we don't lack information about the brain. We live in an era in which information about the brain has seen exponential growth. Rather than having too little information, we have too much and its sheer volume makes it difficult to put the pieces of the puzzle together or to see the bigger picture. What makes the puzzle more complex is that some pieces of the puzzle don't seem to fit – not all information about the brain is good information. Every day, two to three papers are published on some, often obscure, aspect of the brain. Some of the papers draw conclusions that are likely inaccurate or confuse the reader. This mountain of information is a huge challenge for anyone who makes an in-depth study of the brain. To add to this challenge, the boundary between the brain and the mind are not clearly defined.

The premise that is offered here is to create a machine that copies the learning abilities of the brain and evolves intelligence over time. Its method is to evolve intelligent behavior one small module at a time, exposing modules to simple knowledge and then building upon that simple knowledge. Because these modules share a common architecture, they can be combined into larger and more complex synthetic brains.

The "Synthetic Neuro-Anatomy Processor" described in Appendix 1 is not an abstract concept. It is a building block for a synthetic brain that incorporates a large number of emulated neural cells. Like the brain, it learns from whatever environment that it is exposed to through sensors that produce meaningful pulse streams. It interfaces to a microprocessor through a memory-mapped interface, allowing the device to be integrated into a standard PC architecture. This provides a microprocessor with the ability to perform neural processing at speeds that were previously only possible on a supercomputer.

As a final point in this introduction the question is raised, are intelligent machine frightening? According to many Hollywood movies they are murderous fiends, ultimately destroying their creators. People have used trained animals to perform specific tasks over thousands of years. What is the difference then between using a trained animal and using a trained machine?

An elephant or a horse for that matter can do a lot of damage if it is out of control, yet people are willing to accept this risk. While it is true that some people have been killed by trained animals, the benefits outweigh the risks. In many ways we can have more control over intelligent machines than we have over animals. We can turn an intelligent machine off, wipe its memory, and start again. We can indoctrinate a learning machine of sufficient complexity not to hurt people. We determine the type of sensory information that the machine is exposed to. The level of risk is relative to the way the machine is used. A synthetic neural device that is embedded in an artificial cochlea converts audio waveforms into neural pulse trains and injects those into the midbrain. It allows a deaf person who does not have a functioning cochlea to hear. It is not going to take over the person or combine forces to take over the world. In larger machines we can examine and selectively delete training models that we do not like. Because of its modular structure we can start small and build up our knowledge as we increase the size and capabilities of our intelligent machines.

We have nothing to fear from machine intelligence, unless it is a programmed intelligence based on pure logic. Small intelligent devices will be used in dedicated prosthesis, such as pacemakers, artificial limbs, cochlea, and retinas. Dedicated intelligent machines will be used in manufacturing, in safety devices, and in toys. By the time we have learned how to build a machine with the complexity of the human brain we will also have gained sufficient experience to build machines that are safe. In any case, they will be less dangerous than program controlled robots. The primitive industrial robots that are in use in factories today are dangerous machines.

Chapter 1

Artificial Intelligence Reborn

"Any intelligent fool can make things bigger and more complex... It takes a touch of genius - and a lot of courage to move in the opposite direction."

Albert Einstein

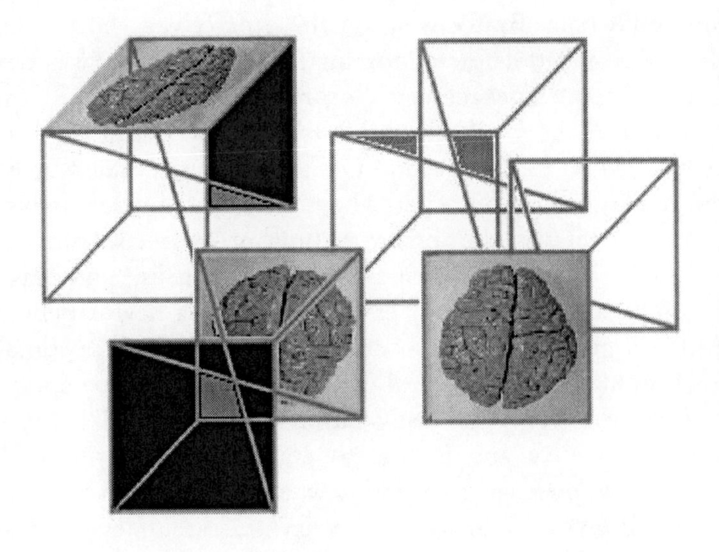

Thinking outside the square

Image created by the Author

A Short History of Artificial Intelligence

A subdivision of Computer Science, Artificial Intelligence was conceived nearly seventy years ago. At its birth it was full of puffed-up pride. Computer scientists had complete confidence in their own creations and had very little knowledge about the mechanisms inside the human brain. That combination lead them to the misconception that they could do better than the human brain, that their creations were far superior and would soon surpass the level of intelligence of humans. Neurons were mistakenly thought of as poorly functioning switches. They argued that the brain was not all that great, that it does poorly in recalling details and that it is extremely slow in performing mathematical tasks. It forgets, or temporary fails to recall important data. By focusing on these perceived shortcomings they overlooked a few important facts. The brain was never intended to be an exact logical or mathematical machine. It was the human brain that gave birth to the technology that eventually led to the invention of the computer, aided by little more than pencil and paper. There are a few obvious points to consider. Firstly, the brain is brilliant at recognizing and identifying complex shapes, sounds, smells and tastes. Secondly, the brain is not easily baffled by a new experience. Rather it gives rise to curiosity. A new experience is a stimulus for learning. It can also deal with new experiences by applying existing knowledge to vastly different situations. Thirdly, the brain is creative and is able to compose great symphonies, combine knowledge to form new ideas, inventions, art, and stories. It gives rise to complex behavior, social interaction, and curiosity and learns from every experience.

A computer may be able to multiply 100-digit numbers in a millionth of a second, but fails entirely on tasks such as cognition, acquiring knowledge by learning, or creativity. A poor replica of some of these actions can be performed programmatically. None of these functions are native to a computer. A computer is a machine that sequentially fetches instructions and data from an information carrier (memory, disk, tape etc.) and performs Boolean logic operations on that data. A simplified form of recognition is created by programmatically comparing stored known values against a set of unknown input values. The computer program works on sequential chunks of data 8, 16, 32 or 64 bits in width (depending on its data-bus width), lacking the ability to process millions of simultaneous events. Therefore it cannot process an entire image, but must cycle through the image comparing the data a chunk at a time.

Creativity requires a level of awareness. No machine has ever exhibited any level of awareness. Computer learning has only recently seen some progress. Dr. Geoffrey E. Hinton[2] has developed a brain inspired method that allows a neural network program to learn. Computer based learning is nowhere near the performance of the learning abilities of the brain. Computer learning is extremely slow and relies on human intervention and vastly simplified images. Current Artificial Intelligence programs have no more than syntactic representations of knowledge[3] (Data) without any form of awareness what the words mean, without context and without the ability to autonomously add to, or to complement this data. This demonstrates an inherent danger. A computer program should never make life-or-death decisions because its decisions are not based on knowledge, awareness or compassion, but only on

---

[2] Hinton, G. E., Osindero, S. and Teh, Y. (2006)A fast learning algorithm for deep belief nets. Neural Computation, **18**, pp 1527–1554.

[3] John Searle. 1980. The Chinese Room Argument

pure logic comparisons that are limited in scope by their binary method and to the programmer's abilities.

A description is not knowledge in itself. This is a very significant difference. Written knowledge is static – and it remains static until someone reads it, thinks about it, associates it with other learned knowledge and applies it to solve a problem. Knowledge is also associative in nature. We associate a word with an object, with a visual image, and perhaps memories of where and how the object is used. Humans have this great gift, called creativity. We can add to the knowledge that we have learned, by experience, by combination of other knowledge, and by invention.

Brain-inspired systems

It has been claimed that computer programs for specific recognition tasks have been inspired by the way the brain works. Such statements are somewhat deceptive. Brain-inspired systems are a long way from 'working like the brain'. For example, a speech recognition system performs a tree structured search instead of sequentially searching through an entire numerical list for matching sound patterns. This is done by creating an index from the sampling of the first phoneme. A list of phonemes is searched until a match is found of all known words that start with that phoneme. The program then jumps to the list that this index refers to. Instead of searching all the words in the entire dictionary, they now only have to search the category of words that match the sound of the first phoneme. This process can be repeated for the second and third phonemes, until the entire sound pattern has been matched. Each phoneme that is matched reduces the list of possible words. Rather than looking for an exact match, the program selects a number of candidates that match, and then selects the highest scoring match from those candidates.

The brain does something similar, but not in the same way. It selects information by selecting a 'best match', eliminating other matches that come close by inhibiting those neurons. The brain's storage has a hierarchical organization. Information is grouped in categories, but the brain does not search and perform a sequential binary compare operation.. The difference may not seem significant in this example where only a few samples and matches are considered. The difference becomes obvious when many millions of samples are involved, particularly where the samples are distorted by background noise. The brain uses input streams to directly address a huge collection of neurons, some of which will respond. Then it selects the best match by inhibiting all other candidates. The result is used to address the next column of stored information, effectively predicting the stimulus that is to follow. The sensory input that follows then confirms that information. This process is performed in parallel, it has better perception in noisy environments, and it is many orders of magnitude faster than searching.

The image shown below shows the difference between sequential searching and direct selection. It may be somewhat reminiscent of old artificial neural networks. Old neural networks are basically correct in structure, but wrong in the method of processing synaptic data and how an output spike is generated.

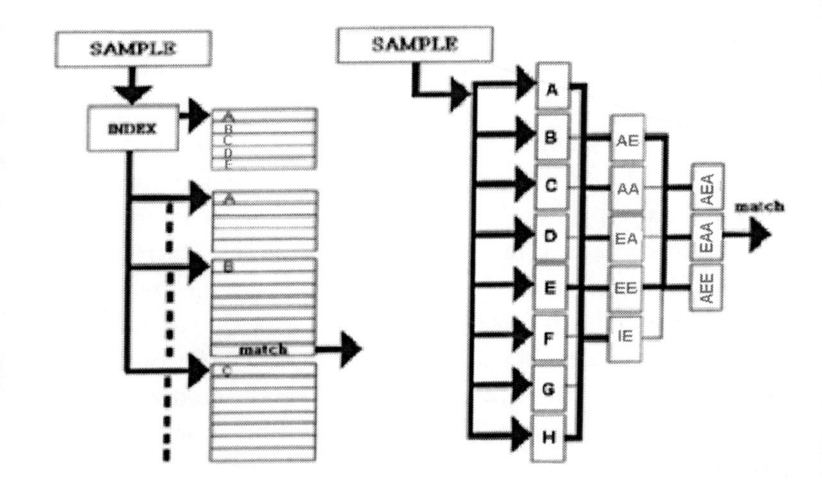

The difference between 'indexed searching' and the brains' method of using the incoming information as the index to address information directly.

Nerves are connected to a set of specific neurons and input pulses are integrated in respect to each other and over time. Time is an important vector, because things in the real world are not only separated by space, but follow each other in a rational sequence.

Another example of a 'brain inspired system' is found in visual processing for face recognition. In a computer the image is digitized and is stored as pixels. A pixel is a dot in the image. Each pixel is stored in memory as a value which consists of intensity and color data. After converting the entire image to data, the position of major facial features is detected by searching the image dot pattern for a stored set of known dot patterns. An eye for instance has a specific fundamental dot pattern. Once their position in the image is known, the software measures the distance between these major facial features. This is thought to be similar to how we recognize people. The methods of extraction and measurement are drastically different in their use of lookup tables, how the program sorts through data, and in how a matching image is determined.

In the brain the image data stream from the optic nerve is used to address information that is stored in synapses in the visual cortex. Specific neurons in the eye have already pre-processed the image to separate color information from the outlines of objects. In V1 these outlines are matched to horizontal, diagonal and vertical line segments. This information is combined into shapes in higher cortical regions. The facial features are determined by saccades – rapid and subconscious movement of the eyes -which provide the brain with timing information, thereby causing the activation of neurons in a specific sequence. The activation sequences and signal timing are typical for a particular face. The brain does not have to learn to recognize a face every time we meet someone new. All faces use the same brain mechanisms. The brain is very efficient in its use of neural modules and easily absorbs variants to existing information. When we meet someone new, the recognition particulars of their face is stored in a structure that recognizes faces, not as an image but as variants.

In both these 'brain inspired' examples, computer programmers compromised the brain method to create something that is workable within a computer environment. A computer's prime function is to calculate, to compute. The brain's prime function is to learn. So a method that consists of matching learned information and neuron timing had to be modified to work by logical comparison and measurement in the computer. Similar compromises are made throughout all narrow Artificial Intelligence (A.I.) programs. Because no standard exists by which to build A.I. programs, every program is built from the ground up. The result is that the A.I. field is fragmented. Because of this fragmentation it is impossible to reuse previous work and it is very difficult or impossible to combine the functions of several different systems.

Purpose

Why do we need smarter A.I. systems? Artificial Intelligence up to now has been dubbed 'Narrow A.I'. This is a term that has been invented to group all recognition and control programs that simulate some facet of human intelligence. Simulating an aspect of intelligence is not the same as intelligence. A speech recognition system does not 'understand' what is said any more than a tape recorder does. Even a sophisticated speech recognition system such as 'Watson' the Jeopardy playing supercomputer just performs logic operations. Digital comparison of sound patterns is far from perfect. Most of us have experienced the unpleasant task of speaking to a machine on the other end of the telephone. The thing just did not get what we wanted and kept coming up with the same annoying list of questions or options. Eventually we got put through to a human operator. The human operator got what we wanted straight away, even if we spoke with some difficulty due to a cold or a sore throat. Why is it that humans can understand one another, and machines are struggling? This is because hearing is much more than pattern recognition. The human brain evaluates the way words are spoken, where the emphasis is and the structure of the sentence. The expectation of the listener is an important factor. If we can see the speaker, facial expressions and body language play a large part. In the same way, intelligence is much more than the logical analysis of data. It involves a capacity for learning, reasoning and awareness. Finally, knowledge is not data. Knowledge is vastly relational, with connections to other knowledge that is not logical but depends on the individual's experiences.

A clear definition of intelligence as well as an understanding of the nature of knowledge is necessary. Without such prerequisites, anything can be (and has been) called 'intelligent' or 'smart'. A typical example is 'smart phones'. The dictionary defines smart as "Having a quick mental ability" or as "A sharp pain". Well, my smart phone has no mental ability whatsoever; it is just running programs. The marketers who coined the phrase 'smart phone' must have had the second meaning in mind then? In a related inclination, to avoid confusion of the old static neural networks with new cognitive technologies, a new terminology has been coined, called "Artificial General Intelligence" or AGI. Even if its purpose has not been realized in the past, the term "Artificial Intelligence" or 'A.I.' describes exactly what its intentions are; an intelligence that is derived by artificial means. A new name is not going to make a difference. A computer program can give the outward appearance to be 'smart', because programmers pour their intelligence into them, but decisions are made on pure logic comparisons. A learning ability is extremely difficult to create within a computer program and does not reflect the learning abilities of a human. Programs are constrained by a rigid Boolean environment that allows only the standard arithmetic functions and comparison of values and binary yes/no decisions.

An existing computer technology that is loosely based on the way the brain recalls information is called 'Associative Memory'. However, its uses are limited because the stored data does not have the relational aspects of stored knowledge. The data is used as the 'address' to recall the data that is stored there. At first glance this does not make sense. Why would we want to recall data from memory that we already have? One useful application is where the available data set is not complete. An incomplete or obscure reference can recall the entire data set. Another application is in data matching, where the method replaces the searching and matching of lists.

In the next chapter we examine an improved method of learning and storing information that is directly based on the way the brain functions, called a Synthetic Neuro-Anatomy.

Chapter 2

Introducing a Synthetic Neuro-Anatomy

"But they are useless; they can only give you answers."

William Fifield (ascribed to Pablo Picasso in Paris Review 32, Summer-fall 1964)

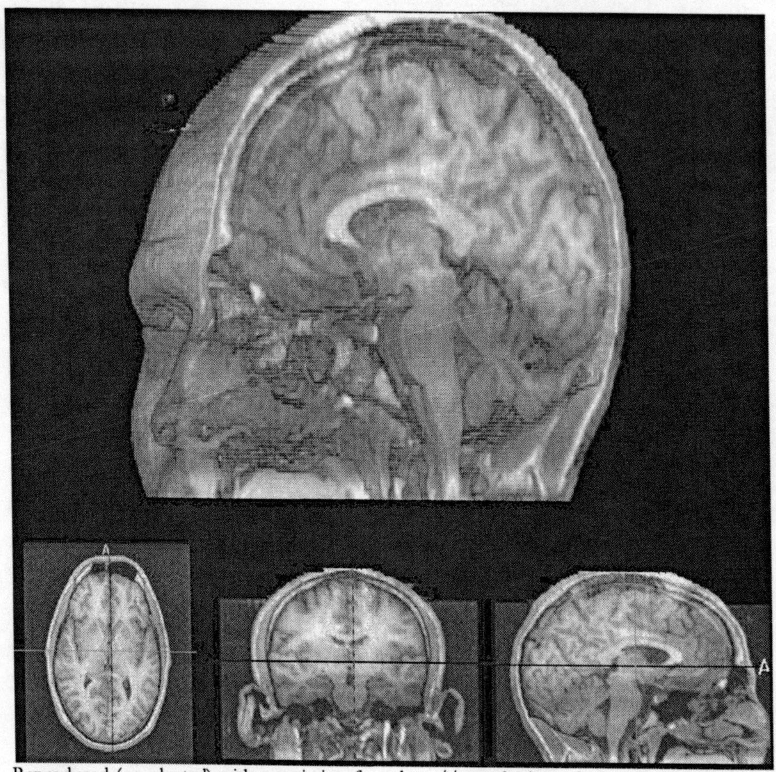

Reproduced (or adapted) with permission from http://www.brains.rad.msu.edu, supported by the US National Science Foundation.

Forget everything you know about computers. The brain works by completely different rules. A Synthetic Neuro-Anatomy is consists of an array of processing nodes that emulate, in hardware, the IO characteristics and functions of a neural cell of the brain. It is constructed from logic gates and learning is its prime function, to store the parameters of relevant pulse patterns as values in registers. Its secondary function is to associate incoming data with previously learned data and to generate an output signal. It has many inputs, but only a single output. The recognition function generates an output signal that reflects the successful association of a previously learned input sequence. A feedback signal confirms this association and modifies values in memory, which is mapped to input signals. A microprocessor interface allows the digital monitoring of all aspects of each neuron's operation. About ten thousand of such single-bit components emulate the actions of a cortical column, and a large number of interconnected columns form a synthetic brain. The number of synapses and neuron density determine the level of intelligence that it can potentially acquire. Its ability to learn complex tasks is directly related to the number of neural nodes, and the total number of synapses in the synthetic brain. The technology is modeled on neuroscience, incorporating the latest findings on the learning mechanisms of the brain. The system learns by repetition and by intensity. The device learns through multi-level feedback mechanisms at node, column, and module level. The memory of events is reinforced with each repetition, but also by the increased intensity of input streams. A synthetic brain module consisting of these components acquires a function that is related to the sensory information that is provided to it. Input streams represent any natural phenomenon that can be represented as a pulse stream from a synthetic sensory organ. Pulse streams can represent the feedback from the movement of a prosthetic limb, or can indicate a specific sound pattern from an artificial cochlear. Pulse streams can represent image information, an odor, taste or the sensing of heat or cold. Like the natural system in the brain, all these different senses are processed in modules that look uniform in structure. Although the structure of these modules is the same, the information that is contained within these modules is not the same. Because the information is different, the nodes respond in a different manner to input streams. The stored knowledge determines the processing algorithm. A Synthetic Neuro-Anatomy processor is an information carrier that is formatted during early life, and continues to improve and learn within that structure.

Differences with current computer technology

The statement at the top of this chapter was wrong to some extent. Computers cannot give you any answers. They can only run programs. A programmer can make a computer give the wrong answer. Programs are large and complex, and can easily conceal many errors and omissions, or intentional inaccuracies. Programs are written by humans, so ultimately it is the human programmer who provides the answer, aided by the binary logic and mathematical functions resident in the computer.

The differences between neural processing, be it synthetic or natural, are huge. A computer is a versatile machine based on logic components that switch between zero and one. Its programs determine if the machine is used for accounting, word processing, as a program development platform, or as a games console. The brain is also a versatile machine, but is based on a large number of interconnected analog nodes. We use our brains for everything from lifting a spoon to eat dinner to performing complex mathematics, ballet, and the creation of art. We recognize words, tastes, images, people and objects instantly, even if part of an object or face is hidden. The brain is an intricate interwoven fabric of cells which store knowledge. It is formatted by learning and never stops learning. It constantly inserts new information at all levels of its knowledge pyramid. In a computer all data has to pass through its central processor. A synthetic neuro-anatomy has no central processor, but consists of thousands of columns, each consisting of ten thousand parallel processing nodes. There is no centralized program. Each node performs its functions independently of any other node. Learning occurs quickly because no 'processor bottleneck' exists.. Each node responds independently to stimuli and generates an output signal that is relative to the spatial distribution and the timing of input signals, modulated by the knowledge that was previously stored in the node. The device can be connected to a microprocessor extending its functionality with neural processing.

Microprocessor compatibility

The entire hierarchy of knowledge that is stored as a training model is available through a microprocessor interface. This enables several interesting possibilities. It makes it possible to copy the 'training model' from a fully trained machine to other machines to provide instant training and innate knowledge. It provides a microprocessor with a large matrix of neural cells, enabling the simulation of brain processes on small desktop computers. This task required large supercomputers up to now.

Training models can also be stored in a software library on a computer, and combined with other training models. For example, the prototype chip was trained to recognize simple tones from a signal generator. A spectrum analyzer was used to simulate the cochlea. The chip quickly learned to recognize ten different tones that were applied to it. The chip later responded to same ten tones embedded within a complex speech pattern. A speech recognition system could build upon this simple application. The training model can be uploaded into a device that has more synaptic memory. Next it can be trained to recognize phonemes, and eventually words and context. The ability to learn complex tasks is directly related to the number of available synapses and nodes Training models for speech recognition, visual recognition, olfactory recognition, etc. can be combined into larger future systems. In this way, a growth path is established that eventually will lead to the emergence of truly intelligent machines. Impurities in silicon no longer matter in the way they do in the production of microprocessors or memory. A faulty node is simply ignored and has no impact on the training pattern or the training time of a synthetic neuro-anatomy brain. Unexpected events do not crash the machine, nor confuse it. If the event repeats it presents a learning opportunity. When 'mind uploading' becomes possible, either through vastly improved MRI (Magnetic Resonance Imaging) scanning or a new, yet unknown technology, then a large structure of these nodes could be used to contain the contents of a living person's brain. Prosthesis equipped with a synthetic neuro-anatomy device could be controlled by the brain like natural limbs, but could also be used in combination with a central Synthetic Neuro-Anatomy brain to construct intelligent robots. There is much yet to discover about this new technology. At this time it is difficult to look over the horizon of its potential. Cognitive processing within a synthetic neuro-anatomy is an exciting new direction in Artificial Intelligence that will evolve over time. It will be some time before we can build a synthetic brain that has the complexity of the human brain. By first building smaller machines for specific tasks we will gain valuable experience on how to construct synthetic brains.

The Reusability of Learned Tasks

At this present time a human-sized artificial brain of 86 Billion artificial neurons and their 100 trillion synapses would fill a stack of 8" silicon wafers 2 feet high. Machines with smaller synthetic brains can perform useful tasks. With a limited number of synapses to contain information, these tasks will need to be simpler than the recognition tasks that a human can perform. Such tasks would be things like olfactory (smell and taste) recognition, visual identification, sound and speech recognition. The technology is also likely to be used in prosthesis such as artificial limbs, artificial cochlear and artificial retina, as well as other brain implants. Each of these devices would have typical training models. Such learned behaviors consist of values across thousands of registers within the synapses of the synthetic brain. Individual values are meaningless. In context they make up a functional set, an elaborate 'fingerprint' of learned behavior, a training model. These values can be read back through the synthetic brain's computer interface and stored in a library file on a computer. These sets can then be reused to create innate knowledge in new machines that perform the same function. They can also be combined in systems that have bigger 'brains'. The training model needs to be restored to synaptic memory, so larger training models require larger numbers of synapses and nodes. By the time the integration level is such that we can construct large synthetic brains, we will also have constructed significant training model libraries for everything from the control of artificial limbs, sound recognition, speech synthesis, vision, olfactory and tactile recognition. Learning never stops, it's an intrinsic feature of the technology. Therefore a synthetic brain that receives innate knowledge from a library file continues to add to the knowledge. This creates increasingly smarter A.I. systems, building upon the experience and training of previous machines. A synthetic brain that works along the principles of a biological brain learns rather than being programmed, and allows increasingly intelligent machines to evolve. Each generation building upon and re-using the knowledge that was learned by a previous generation of machines. This is an incremental growth path in contrast to the 'singularity', which is a tentative future event, an extremely fast supercomputer that runs a program that is 'smarter than people'. However, it is not clear what this means. A pocket calculator performs arithmetic at least a thousand times faster than a human, but is it therefore 'smarter than people'. In the context of "more intelligent than people" there is little hope for digital computers. All past attempts to reinvent intelligent behavior using a computer have failed.

Applying Learned Knowledge

The same mechanism that stores knowledge in the synthetic brain is used to recall information. The input-streams connect straight into synaptic memory and recall the knowledge that is stored there. The neuron body performs a spatial temporal integration of the addressed synaptic memory content to determine if the match is adequate.

The brain uses analog values that are stored in synapses to match the timing of pulses, not only by their spatial distribution (its distribution of ones and zeros) but also how pulses exists in time and in relationship to the timing of other pulses that were received. The stored values represent a pulse pattern indicative of a previous event (training) that is recalled by association.

Like the brain, the synthetic brain uses thousands of synaptic values to identify timing and pattern features in the pulse streams that are coming from sensory devices or previous layers of neurons. These neural nodes are organized in a hierarchical memory structure. In the first layer simple features are identified, and in the next layer combinations of features are identified, and so on to eventually match complex entities such as shapes, words, sounds, movement etc. A minicolumn is a brain structure that is able to match complex features. A minicolumn consists of approximately 10,000 neural nodes, with each node consisting of synapses, neural and glial cells. Sensory input streams are not random 'noise', they have a predictable structure; actions follow each other in a rational chronological order. Therefore the next sensory input can be anticipated and then confirmed by input streams. There are mechanisms in the brain's columnar organization that match this predictive behavior. The next layer of columns is preempted before sensory input arrives.

The brain and the computer

Reverse engineering the brain was started in earnest in 2003 by examining the available literature. There is a lot that has been written about the brain, mainly over the last 50 years. Research papers frequently contradict each other at some point. The human brain is superior to any man-made supercomputer. It is a 3D 'wetware' storage facility with a component density that is higher than any computer chip. It can store a lifetime of experiences and memories. It accomplishes this amazing feat at a power consumption of less than 20 watts and a tiny volume of around 1400 cubic centimeters. A human brain will comfortably fit into a man's hand with the edge of the fingers spread around the outside. It is not as big as most of us would like to believe it is. It feels heavy for its size, between 1.35 and

1.4 kilograms. A supercomputer, in contrast, is a behemoth that consumes 6,000,000 watts and occupies the area of a basketball court, yet falls short of the processing power of the brain. The brain gets upgraded every day, increasing its memory capacity by growing new storage locations on demand as you go about your business. Information is stored in a way that it is instantly accessible. We don't need to think about how to drive our car, or how to walk. Conceptual knowledge requires but a small clue to trigger the recall of the entire set. For instance, we recall much of a conversation once we get a few hints. This is because of the hierarchical structure of our brain. We need information to associate to get the chain of recall started. We need an entry point which can link anywhere in the pyramid structure. The neurons produce varying bursts of electrical impulses when the synapses contain a memory that matches incoming information streams. Each neuron is a processing node that stores information in thousands of attached synapses. Some specialized neurons have as many as 200,000 register connections (synapses). Most of this complexity evolves as we learn from day to day. At birth, only the most fundamental connection plan exists, pre-arranged by DNA. This includes the control of automatic body functions, basic feeding responses, and the general layout of specific modules and connections between modules and sensory organs. There are also many millions of random connections that are removed in the first year of life. The 'connectome', the grander connection plan of a fully developed brain, evolves from day by day experiences and is as individual as a fingerprint. The connectome of identical twins is different. We can remember many thousands of facts by the time we are five years old, besides all our other accomplishments, such as learning to control our limbs, stand up, walk and acquire a language. From the moment we are formed our brain increases in complexity. Over the duration of our lifetime, our brain stores many millions of facts, directions, information, languages, movements and academic knowledge. Our object recognition skills are astonishing. We instantly recognize people, places and things no matter how obscure they are. We can figure out most people's scribbles no matter how bad the handwriting is. As the old adage goes, we are not likely to forget how to ride a bike, or catch a ball. As soon as we read the first page of a book we not only remember if we have read it before, but also what the book is about. We remember the words and melodies of songs, the movies we have seen and places we have been to. A smell or a taste can take us instantly back into a memory. The brain also has remarkable built-in redundancy. Millions of neurons can be lost, and the brain carries on as if nothing happened. The brain has a way of repairing itself even when a stroke or an accident causes major damage. Neural plasticity – the ability of brain cells to be reused for another purpose – eventually reroutes neural signals around a damaged area. The brain has a high density of neurons; the average is 61 million neurons per gram of brain matter (86 billion divided by 1400 grams). But what is really staggering is that there are over 70 billion synapses per gram, and each one is a multi-level storage location. This is a higher density of storage components than we can produce in a silicon computer chip. A gram of brain matter holds some 70 Gigabytes of relational information and 61 million processing nodes. This cannot be directly related to computer storage. The organization of information in the brain is such that each storage element is used in many different contexts. Therefore the sum is greater than the total number of storage elements.

Every second of every day more information is added to our mind. There are things that we do not even know that we remember, things that happened a long time ago and trigger emotions without us really knowing what it is all about. We have a conscious, a subconscious and an unconscious mind residing in our brain. The information stored in our brain gives us skills that enable us to do our job, sports, compose music, create art and write books. We have imagination and dreams. Our brain warns us of imminent danger and makes instant estimates of distance, speed and time. Most important of all it makes us who we are. Our brain is where our mind and our personality reside. Even if we had the technology to transfer our mind, none of us would want to trade our mind for the perfect recall and the cold organization of a computer. We only start realizing how awesome our brain is when something goes wrong. I experienced the heartbreaking change of personality and decline in my father as the effects of Alzheimer's disease took hold. He slowly deteriorated over a period of six years. What had been a proud, gentle, well-educated and intelligent man became a stubborn and irritable bully, and eventually a vegetable before death set him free.

We all know that people are imperfect. We forget things, misplace our keys and sometimes even large things like our car in a parking lot. At times we are left wondering who was driving when we do not recall the past 5 kilometers, or we find ourselves unintentionally driving to work on a Saturday while our brain is preoccupied with something else. We experience embarrassing situations when we remember a face but we cannot put a name to it. Sometimes this even happens with people we have known for years. Our male brain gets 'overloaded' when we try to do three or more things at the same time, but females seem to have no problems with that. Our brain is not particularly fast or accurate in mathematics. A computer easily surpasses the brain's abilities in data retention and mathematics. But in cognition, creativity, analytical thinking, invention and the ability to cope with unexpected situations the brain reigns superior. In all its perceived imperfection, the brain is a magnificent creation.

STDP – BCM Learning

STDP stands for Synaptic Time Dependent Plasticity and it is one of the theories that describe the learning process of the brain. BCM learning, named for named for Elie Bienenstock, Leon Cooper, and Paul Munro, is another theory. BCM learning results from pulse intensity while STDP results from the repetition of a defined pulse pattern.

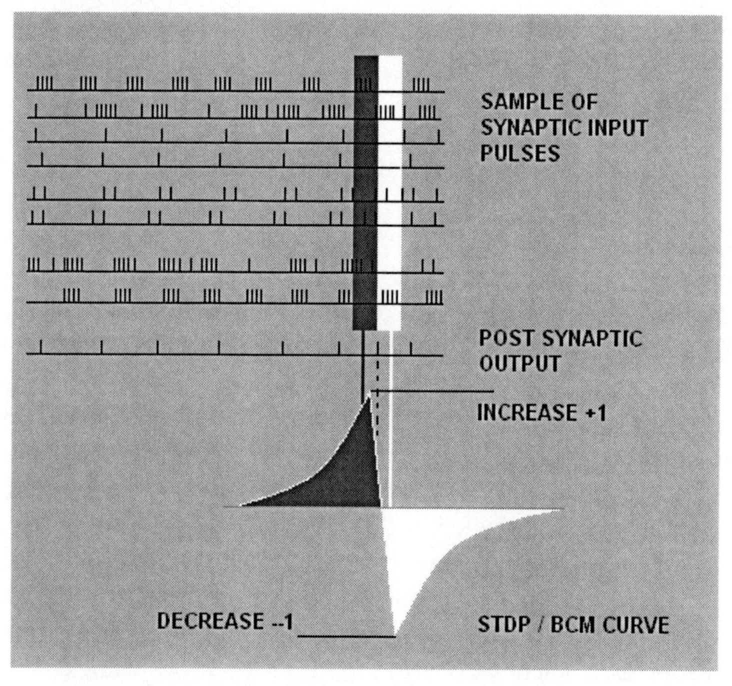

STDP /BCM learning method

As a particular set of pulses trigger a response from the Post-Synaptic integrator, the synapses that contributed to this event are strengthened thereby increasing the likelihood that in the future a similar pattern will lead to an output event.

Integration of the two methods has resulted in a method that learns by intensity and repetition, and is closer to the way the brain learns. An input pulse causes neurotransmitters to be released from the synapse into a synaptic cleft, which is a small gap between the cell body and the synapse bud. The cell body contains receptors for specific neurotransmitters. The cell's membrane potential is changed temporarily when a receptor absorbs a neurotransmitter. It increases for excitatory neurotransmitters and decreases in the case of an inhibiting neurotransmitter. The neurotransmitter is then taken back up. The electro-chemical effect decreases over time. The rate of this decrease depends on the neurotransmitter type. The type of neurotransmitter is determined by the previous neuron, which synthesizes the neurotransmitter and passes it to the synapse node. The integrated value is time dependent or "Temporal". The membrane potential of the cell is a function of the spatial distribution of inputs, the pattern of pulses in relationship to one another. A single synapse that is pulsed intensely will cause the synaptic cleft to saturate and to deplete the neurotransmitter level of the synapse. Feedback returns the effect of an output pulse of the post-synaptic neuron back to the synapses. The synapse response is modified depending on the time difference between input and output pulses. The synthetic version of a neuron consists of many identical synapse circuits. Feedback, a temporal integrator and a pulse shaper complete the neuron circuit. In this way each synapse in the set is trained to respond to a specific combination of pulses distributed spatially and over time. The neuron body integrates the synapse outputs and responds strongest to a single pattern. The output rate and the intensity are directly proportional to the level of coincidence of the input pattern. A large matrix of such devices is able to learn to respond to complex patterns. In the synthetic neuro-anatomy the neuron nodes are organized in mini-columns, in the same way that such columns exist in the brain [4]. The registers in a column can be mapped to a microprocessor as memory, so that the training model stored in these registers can be copied into a training function library. Training patterns can also be restored through this interface, making it possible to seed the matrix with a training model, on which the system then continues to build as it continues to learn. This new technology makes real-time neuro-computing a reality. Millions of neurons and hundreds of millions synaptic circuits can be integrated on a single wafer. Because of the slow switching speed of these devices they remain cool and the supply current is low, thus there is no need for forced cooling. Faulty circuits caused by impurities in the silicon wafer are simply ignored when the system builds its processing centers. Stacks of wafers could be used to start approaching the complexity of the human brain. A human mind, once scanned, could be uploaded into such a machine. A synthetic brain like this will require artificial senses, or it will exist in darkness and silence. Lacking any stimulus, feeling or excitement it will go quietly and utterly insane. A brain needs to be stimulated to function. The mind, and intelligence, does not evolve without stimulus. In the next chapter we will examine what is required to create an intelligent entity, and why all previous attempts have failed.

---

[4] Vernon Mountcastle

Chapter 3.

Defining Intelligent Systems

"A Fox entered the house of an actor and, rummaging through all his properties, came upon a Mask, an admirable imitation of a human head. He placed his paws on it and said, "What a beautiful head! Yet it is of no value, as it entirely lacks brains."

Aesop (Ancient Greek storyteller, 620-564 BC)

Aesop

In 1962, Rosie appeared as the servant robot in the kids cartoon 'The Jetsons'[5]. She did all the housework, looked after the kids and she even made a sarcastic comment or two to the man of the house. In the 1977 release of the first Star Wars[6] movie, a robot called C3PO not only conducted intelligent conversations, but showed determination and intent. Alas, today Rosie and C3PO are still mere fantasies. Intelligent robots do not exist. Today's robots are awkward machines in factories controlled by a computer program. The label 'robot' is also used for machines that are remotely controlled by a human, such as bomb disposal machines. Over 65 years of research, simulations and building models has not resulted in anything close to intelligent, autonomous robots like Rosie and C3PO. Some people may disagree with this statement, and point at a number of experimental and 'home' robots. I would argue that these are not intelligent, and that classifying them as such is based on a poor understanding of intelligence. These experimental robots are machines equipped with sensors that are controlled by a computer program and repeat the same steps for the same set of stimuli.

Building Intelligence

For many years, people have anticipated building an intelligent system. Since the creation of the electronic stored program computer in the 1940's, people have attempted to create artificial intelligence on computing platforms. In 1822 an Englishman called Charles Babbage

[5] The Jetsons © 1962 Hanna Barbera

[6] Star Wars, © 1977 Lucas film. George Lucas

designed a machine that he called the Difference Engine. Babbage knew that the mathematics in his machine was performed by mechanical wheels and cogs. When the question was raised, Babbage noticeably denied that the machine was intelligent. For some reason, a machine that performs mathematics speaks to the imagination. A human is intelligent and can perform mathematics, but that this does not mean that a machine that performs mathematics is intelligent. Over the next ten years Babbage worked on an even more ambitious project called the Analytical Engine. Had this machine ever been built it would have been the first programmable computer. The Analytical Engine was a mechanical, steam-driven colossus the size of a locomotive. About a century later, around the time of the 2nd world war, both Alan Turing and John von Neumann philosophized about intelligent machines. Von Neumann wrote his last book "The Computer and the Brain" on his deathbed. Alan Turing pondered the question whether machines could think in his 1950 paper "Computing Machinery and Intelligence". Both men considered the entire, fully developed adult mind rather than a hardware platform with a software layer. They considered the brain to be a control system, rather than a dynamic learning system, a common mistake that has persisted. Both these men assumed that computers would soon evolve to a level and speed to overtake human intelligence. They did not have access to the knowledge about the brain that we have now and they underestimated human intelligence. Developments in the computer industry over the last 65 years have proved that faster and more advanced control systems does not lead to those systems being intelligent. Control systems have no awareness, cannot learn and cannot build upon knowledge, and lack the ability to combine knowledge to apply it to different circumstances. They therefore miss all the attributes that define an intelligent system. They just run programs.

Evolving Intelligence

We consider that a fully developed adult human brain controls our skills and our body. But how did it get there? A baby is quite helpless. Even though it has spent nine months developing it is not born into the world with complete knowledge. DNA programs the brain to a level that enables it to learn from sensory input, it does not contain enough bits of information to store all our parents knowledge into the infant brain. A baby starts learning before it is born, from the time that the brain is formed. In the womb it moves about, and the brain receives feedback from its limbs. A baby may hear some muffled sounds from the outside world, but the womb is much like a sensory deprivation chamber. Immediately after birth all the senses start to provide huge amounts of information. Its brain starts learning at an increased rate. The eyes provide images of the world and the brain starts to format its visual recognition system. The skin provides tactile information about its environment. The muscles and tendons in the limbs provide feedback about their position and tension. This feedback modifies the neurons in the motor areas of the brain. Each movement refines the stored data and eventually the limbs can be controlled. The learning process does not stop, it continues until the brain stops functioning at death. After the initial reorganization of the motor cortex, it continues to learn. Complex movement control is built upon simpler movement control in a hierarchy. Such a hierarchy is found in all sensory processing centers of the brain. Therefore we can learn new skills and cope with unexpected events. We use existing knowledge hierarchies to deal with the unexpected, learn from the event which updates, and insert new information into the existing hierarchies. This perspective considers the brain to be an information carrier and a learning system that creates its own hierarchical memory structure. New neurons and connections grow where they are required. During sleep the brain consolidates its memory structures. Sensory perception is repressed, and learning from sensory input is suspended. A synthetic brain copies the parallel organization of the brain as well as it method of learning, gathering and storing information. The following example clearly illustrates the contrast between “fast” sequential computers and the massive parallel structure of the brain; Dr. Eugene Izhikevich[7] of the Neurosciences Institute at the University of California, San Diego, simulated a large matrix of simple neurons in software on a computer network. One second of brain time in an array of 100 billion neurons were simulated. A networked cluster of 27 computers took 50 days to complete this simulation, totaling 27 x 50 x 24 x 3600 = 116,640,000 seconds. Considering that this was a simplified, and not a realistic model of the brain, it is to be expected that the brain is at least 120 million times faster than the average state-of-the-art PC. This is due to the brain’s massive parallel organization and vastly different processing method. The speed of a single neural node is rather slow, taking several milliseconds from a change in input to produce a change in output signal. The brain is perfectly matched and synchronized with the world around it. It has to be, in order to learn from it.

Open Questions

Even though there are still a number of open questions, we know a lot about the brain. New research tools, such as fMRI (functional Magnetic Resonance Imaging), CAT and PET scans have helped to expand our knowledge in recent years. Reverse engineering the brain has proved to be a challenging task, not

---

[7] Dr. Eugene Izhikevich. “Why did I do that?” 2006

because of lack of information, but because of the enormity of it. Adding to the confusion is a poor understanding of the division between brain 'software' and 'hardware'. Several brain research methods are discussed here, as well as the details of the brain and what that means for the development of Artificial Intelligence.

By building synthetic brain systems, the more specific questions that we hope to answer are:

- A small synthetic brain has been observed to learn from sensory input, but will a large functional emulation of 86 billion neurons develop human-level intelligence?
- Will a synthetic brain forget?
- Will it need to sleep?
- How much of the human brain will we need to copy? A parrot brain is much smaller but it can produce complex behaviors including decision-making, counting, color recognition and complex speech.
- The connectome is the complete connection diagram of the adult brain. It is unique to every human. Will it evolve as expected once basic connections have been put into place?
- Once we have attained functional brain emulation, will we be able to upload a human mind into the machine?

The only way we can get answers to these questions is to start building machines. It would stand to reason that an accurate digital emulation of the brain would function like a biological brain. At this time we do not yet know what an accurate emulation is. The way Edison developed the light globe was to build one, learn from it what he was doing wrong, and build another one. He repeated this process until he got one that worked. Then he and others improved on that design. The method to develop an accurate synthetic brain may not be the same, but in a similar way we learn from mistakes. Computer simulations are far too slow, and too limited by the speed and sequential nature of a computer to act as a real, learning and interacting brain. No computer simulation can come close to the performance of neuro-morphic machines. By building prototypes we learn from the unexpected, while computer simulations often only confirm what we already know.

Chapter 4

Copying Intelligence As We Know It

"Instead of trying to produce a programme to simulate the adult mind, why not rather try to produce one which simulates the child's? If this were then subjected to an appropriate course of education one would obtain the adult brain"

ALAN TURING, 1950 "Computing Machinery and Intelligence". Mind 49 pg. 433-460

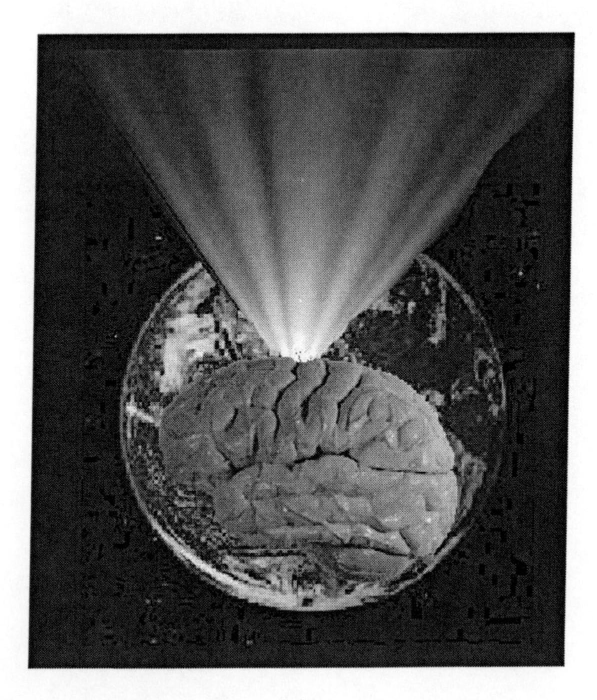

The Definition of Intelligence

To understand Artificial Intelligence we will have to define what intelligence is. We need a workable definition of 'Intelligence' to identify how we are going to build intelligent systems. Marvin Minsky, founder of the MIT Artificial Intelligence Laboratory said that "Artificial Intelligence is the science of making machines do things that would require intelligence if done by men". At first glance this makes sense, but when we consider a few simple examples it appears to be a confusing statement. Adding two numbers together requires intelligence if done by humans, but does that make a pocket calculator intelligent? Even a thermostat could be classified as 'intelligent' by this definition. Would it not require human intellect, and an accurate thermometer, to keep the temperature in a room constant? While it is obvious that Minsky did not intend his words to be applied in this way, his statement does nothing to define intelligence.

Intelligence goes much further than controlling a process or providing a solution to a specific problem. A great work of art is not a solution, but it is a result of the accumulated knowledge and talent, the artistic intelligence of the artist. Someone who lives on the land has accumulated knowledge what soils are best suited to grow certain plants, whether they grow in full sunlight or in the shade, and how much water and what nutrients they need. He may experiment with these variables and learn more, increasing his agricultural intelligence.

I.Q. tests do not really test intelligence, but mostly academic ability. What is intelligence then? There is no simple answer. An approximate answer to this question could be that it is the ability to accumulate knowledge, and the sum of accrued knowledge which is applied to make decisions that benefit the person and others. Intelligence always presumes a level of awareness.

Too complex or too imperfect to emulate?

Some people have decided that the brain is too complex to understand, that the brain cannot understand itself. To quote James and Jean M. Goodwin:

'Despite many assertions to the contrary, the brain is not 'like a computer'. Yes, the brain has many electrical connections, just like a computer. But at each point in a computer, only a binary decision can be made—yes or no, on or off, 0 or 1. Each point in the brain, each brain cell, contains all the genetic information necessary to reproduce the entire organism. A brain cell is not a switch. It has a memory; it can be subtle. Each brain cell is like a computer. The brain is like a hundred billion computers all connected together. It is impossible to understand because it is too complex.'[8]

Others see the brain as too 'imperfect' to copy and believe that we can do better through programming to create cause-and-effect machines that use exact data. Among this group are a number of prominent robotics institutions and A.I. researchers that have tried and failed to reproduce some aspects of the brain. Even so, there are worldwide pursuits going on to understand the processes of human intelligence and to develop intelligent 'thinking' machines. A fully evolved adult brain may be too complex to understand, but the processes that create this complexity are not. It is likely that intelligence will evolve in time when we create machines that are able to faithfully reproduce these processes..

---

[8] Jean M. Goodwin (b. 1946), U.S. physician, professor of psychiatry, and James S. Goodwin. "Impossibility in Medicine", The Nature of the Impossible, W.H. Freeman (1987).

Understanding the Brain's Processing Method

It is important to make a clear distinction between the brain's 'software' and its 'hardware', between its 'programs' and its 'processor'. This is not as easy as it seems. The hardware is a huge matrix of interconnected glia and neural cells, covered in synapses. These are organized in columns, each containing about 10,000 neural cells, a million synapses and 100,000 glia. The brain's learning algorithm creates new synaptic 'registers' whenever they are required and constantly updates the values that are stored in other synaptic registers.

Taking a holistic approach to analyzing the brain with all its accumulated knowledge is analogous to analyzing the processing method of a Windows based PC by looking at the operating criteria of a few transistors at a time. For instance, say we carefully attach some wires to its circuitry and analyze the signals while we make a program perform some action. We will see different waveforms depending on the actions performed by the program. If our sensor wires are on a transistor in the data bus circuits we will see intense waveforms when the computer is accessing a large data array. On the other end of the spectrum we can experiment with the computer's software and analyze its behavior. Will this teach us how the computer works? Of course not. Because the circuitry of a computer is well documented, a computer scientist already knows what to look for, and can quickly identify the data and address busses, control signals and I/O. The brain is not that simple or well defined.

Applying the same principle to the brain

Neurologists attach wires to the brain of animals and record the input and output waveforms of single neurons. They get the animal to perform some action and get really excited when they see a particular waveform, which is considered indicative of that action. Like the computer example given above, without a thorough understanding of the brain circuit there is little chance that individual signals will tell us anything useful. Trying to analyze circuitry from this perspective only leads to confusion. There is an added level of complication. The brain's processing algorithms are constantly modified by the knowledge that is stored there. This information is modified every time it is accessed. Because of the changing nature, the plasticity of the brain, any image is a snapshot of a constantly changing system. Even though the interaction between individual cells is relatively simple, the whole system is inconceivably complex, far more complex than any computer circuit. Interaction between cells consists of sequences of electrical pulses. These electrical pulses are generated when the input streams to the neuron contain information that matches previously stored values. Values are stored in synapses while we are learning. Learning takes place through feedback. Feedback occurs between cells, layers of cells and columns, but also through feedback from the senses and the environment. This whole fabric makes an enormous interactive information carrier that has near instant recall, and it is capable of storing a lifetime of experiences in minute detail.

An intelligent system design based on neuro-Anatomy

Creating an intelligent system design along these lines has required a new way of thinking; to let go of the concepts of traditional 'program control', to let go of a central processing unit that operates by logic and separate memory that stores exact data. Synapses are much more than connections between neural cells. They are an electro-chemical memory that stores a chemical marker value. This value is recalled when the synapse receives an input pulse. The value is updated depending on the actions that follow. The number of connections to a cell, and therefore the number of synapses on a cell is directly related to the amount of information stored in that cell. Input pulse streams address a large number of such stored values, which are then integrated in the neuron cell body. The output of the cell is directly related to the matching level in time and spatial distribution between the input pattern and the stored pattern. The column structure is such that the most prominent cell (the closest match) wins the contest. Routing through the matrix of cells is determined by synapse content. Because of the way this fabric of cells is constructed, the processing algorithms are modified in subtle ways every time synaptic values are referenced. Memories are made up of many millions of those values. Memories are stored as linked lists in columns of neural cells, each column triggering the next column in anticipation of sensory input. Brain arrangement matches the arrangement of the world. The world is a place in which one event leads to another in a sensible sequence. In the same way information in the brain is stored as sequential lists. New synapses form as we learn, inserting new facts and links within existing knowledge hierarchies. Our intelligence is the cumulative effect of all we have learned from the moment our brain was formed. Intelligence is not control and cannot be programmed. It must evolve. Isaac Asimov's[9] 'Three Laws' are impractical from

---

[9] Isaac Asimov "The three laws of robotics"

A robot;

1. may not injure a human being or, through inaction, allow a human being to come to harm.
2. must obey the orders given to it by human beings, except where such orders would conflict with the First Law.
3. must protect its own existence as long as such protection does not conflict with the First or Second Laws.

several perspectives. They are stated in natural language, not logic. Therefore it is not viable to program them into a computer, and we cannot hard-wire them into a freely learning machine. Controlling the sensory input stream can influence what the system learns, but such a restriction will almost certainly limit the development of intelligence. For instance, if a child is exposed to meaningless sounds early in life, then it never develops a proper processing center for speech. Because speech and words are such a big part of our thinking, many other parts of the functioning of the brain are affected[10]. Creating a system design along these lines has required a new way of thinking; to let go of the concepts of traditional 'program control', and of a traditional computer structure. The information carrier layer consists of a massive parallel fabric of emulated cells, organized as a cortical minicolumn. The software consists of the learned 'knowledge' that is stored in registers within those cells. The system self-programs as it learns. Data is not retrieved through an address bus, as it is in a computer, but by association. Hence, there is no address bus, but a number of massive data busses. The number of connections to an emulated cell is directly related to the amount of data stored in that cell. Routing through the matrix of cells is determined by synapse content. The processing algorithm evolves, is complex and is subtly modified every time synaptic values are referenced. Memories are made up of many millions synaptic values. New synapses insert new facts and links within existing knowledge structures. The devices learn from sensory input data and autonomously form processing centers as connections form, in much the same way that a baby's brain evolves and forms processing centers. The famous

---

10 "On the Brain". Prof. Michael Merzenich. University of California, San Francisco.

'three laws of robotics'[11] do not, and cannot apply, because there is no program to control what the devices learn. Learning is an intrinsic function of the design.

Narrow A.I.

Narrow A.I. is defined as a program that solves a specific problem. Its scope is limited to that one problem. Inevitable narrow A.I. is a programmed control system. Computer programming is a product of the human programmer's intelligence. The resulting program is executed on a sequential Boolean machine, which imposes limits on the expression of the programmer's intelligence. Boolean machines work by strict yes/no, 1/0 or true/false decisions. In contrast, accumulated knowledge forms the building blocks for intelligence. Knowledge, beliefs, true and false assumptions and illogical associations govern the outcome of our decisions. By its very nature, knowledge, and therefore intelligence is never exact or logical. This is not a shortcoming of our brain, but a benefit. Because of this inexact nature, we are creative and we see patterns that defy logical analysis.

Narrow Artificial Intelligence has little to do with machine intelligence. What is currently known as narrow A.I. should just be defined as control systems. It covers everything from electronic controllers, automatic sorting systems up to various forms of robots. Branches of this technology include chat bots, diagnostic systems, speech recognition and speaker identification systems, image recognition, navigation, and goal-seeking programs. Narrow A.I. has little to do with the way the brain works. A chat bot is a computer programmed to interact with a human through speech recognition and synthesis. A

---

[11] Isaac Asimov. 1940

casual observer may be deceived that the system is 'intelligent'. The bot responds to human speech with preprogrammed responses, like a tape recorder that has instant access to a large collection of phrase. All actions of the system are executed under program control, and the system does not have any form of awareness. Many of us have had the experience of talking to machines on the phone, machines that did not 'get' what we said no matter how well we spelled it out. In fact, spelling it out confuses the machine completely.

Alan Turing (1912-1954), an English computer scientist, believed that computers of his time were many times faster than our brains. On purely mathematical ability he was right. Our brains were not built to perform fast calculations and a simple pocket calculator is faster. Alan Turing's point of view has been proven wrong in respect to cognitive functions. With computers that are thousands of times faster than computers in Alan Turing's time, A.I. technology does still not even come close to the reasoning, recognition, creativity, and analytical abilities of the human brain.

Industrial robots are controlled by a computer program. They are clumsy and dangerous machines that can only operate in a predictable and simplified environment. They repeat the same actions over and over again. It is very likely that a human would be killed or seriously hurt if they were to stray onto the assembly line. If you would walk up to it, an industrial robot in a car assembly plant would try to weld a hinge to your ear. Reprogramming a robotic production line is a cumbersome operation that can take many months. Industrial assembly robots are very different machines from the simulated human robots of science fiction. There is much tinkering going on in A.I laboratories at universities and other institutions around the world. A robot control program may take, for instance, input from vision sensors and detect the presence of a face. The robot is then programmed to turn its attention to that face. It does this by moving its eyeballs towards the face, giving the impression that it is listening. Such forms of control are not intelligence but it is easy to see how the public may mistake it for the real thing.

Neural networks are another branch of computer science that notwithstanding their name, have little to do with the way biological networks operate in the brain. They date back to the 1940's when the first models of brain cells were built by McCulloch and Pitts. Between 1957 and 1962 [12], misunderstanding the function of a neuron, Frank Rosenblatt devised the Perceptron Neural Network. The Perceptron consists of a matrix of nodes that use a weight value to determine a path through the matrix. By 1969 Marvin Minsky and Seymour Papert [13] proved mathematically that such Artificial Neural Networks (ANNs) could not do very much beyond simple data classification tasks. A few determined individuals have attempted to disprove this point by adding to the Artificial Neuron Network concept. Feedback paths and complex methods have been added. These are not grounded in the biology of the brain at all.

A device that is derived from brain function is a platform technology that has no defined function without learning a skill. A microprocessor is a platform technology; it does nothing without a program that determines its function. An infant's brain is a "blank slate". Will he or she create great art,

---

[12] F. Rosenblatt "The Perceptron, A perceiving and recognizing automation" 1957 Cornell Aeronautical Laboratory

[13] Perceptrons. By Marvin Minsky and Seymour Papert. Published in 1969.

be a doctor, a farmer, a mathematician, a truck driver or a teacher? It depends on the child's natural ability, the environment that it grows up in, its early childhood stimuli and the type of education that it receives. The brain is a 'programmable' system. It is programmed with the knowledge that it receives. This knowledge includes the basic skills like the means by which we move our arm, or move our legs to walk. Its use is flexible. We can change careers or interests at any time in life. As circumstances change, we adapt. That is one of the qualifiers of intelligence – the alibility to adapt to new circumstances.

Why is it that we have not been able to build an intelligent robot like Rosie or C3PO? Is it because we keep building on existing systems and add to the errors in the original concept by introducing new concepts that are not based on biology? The human brain is a remarkable example of an intelligent entity, yet we largely ignore its architecture and methods.

To build an intelligent machine we need to have a sound definition of what intelligence is, what the structure of knowledge is, how knowledge is stored and retrieved by association, and how we learn. Animal brain research falls short in its scope of testing small sections or individual signals of the brain. The structure of an animal brain has significant differences to a human brain. Animals are intelligent to a limited degree – they have the ability to learn and to adapt, but they have no language to reason things out like we do.

Chapter 5

Designing for Intelligence

"If the Aborigine drafted an I.Q. test, all of Western civilization would presumably flunk it." ~ Stanley Garn

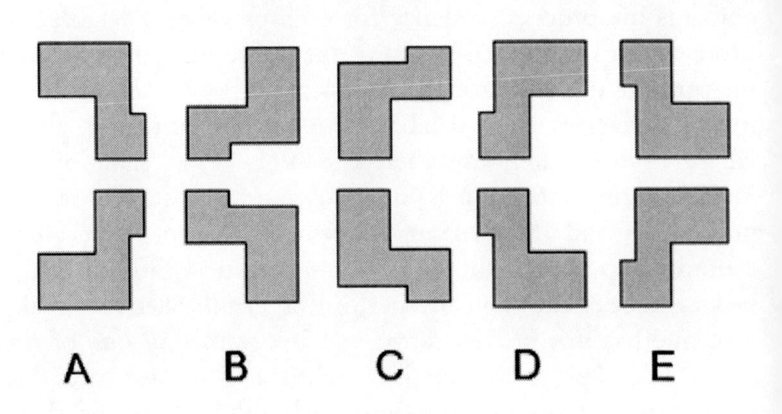

A to E are mirrored images. Which set can not be rotated to reflect all of the other sets? Image by the author

Designing A.I. systems has often been a matter of trying to fit a square peg in a round hole. System design begins with the end result in mind. At the start of the project a detailed specification is drafted. The specification defines in great detail what the system must be capable of, its operating criteria, how it is going to be built, and how long each milestone and therefore the overall project are going to take. Let's take an example from a typical 'narrow' A.I. project. Narrow A.I solves a specific problem, in this case picking robots in a warehousing facility. These robots move cheese to shelves for aging and into processing plants at different stages of manufacturing. After all the stages of the process are completed they move the cheese into the appropriate truck at the loading dock. The controller board on each picking robot communicates over a wireless network with a central computer. This central computer contains the process sequence for each product, a database of products on shelves and a map of the warehouse. It also tracks the position of all the robots and keeps track of orders that are to be delivered. The database contains information about where in the warehouse each cheese is, what state of the manufacturing process it is in, the type of cheeses that are in production and the ordering information from clients. The central computer also needs to communicate with each of the picking robots and regularly perform a 'health check' to make sure one has not broken down and is obstructing one of the aisles. Local programming on each robot makes sure that picking robots do not run into each other. Each of these picking robots looks like a heavy forklift truck. It would be dangerous if a malfunctioning machine were to drive through the warehouse wall into traffic. A failsafe is imposed that cuts power to the machine in case of such a serious malfunction. A variety of sensor devices provide the controller boards with data about the speed, the position, the orientation, the distance to the shelf, and the presence of a load. The wireless network data flow includes the coordinates of each robot in the warehouse, the task of each robot, and its progress. Local data in each robot controller includes its location in the warehouse, distance to the shelf, and alignment of the cheese pallet guides, the vicinity of other picking robots and the velocity of nearby robots. The actions of each picking robot and of the central computer are controlled by a program.

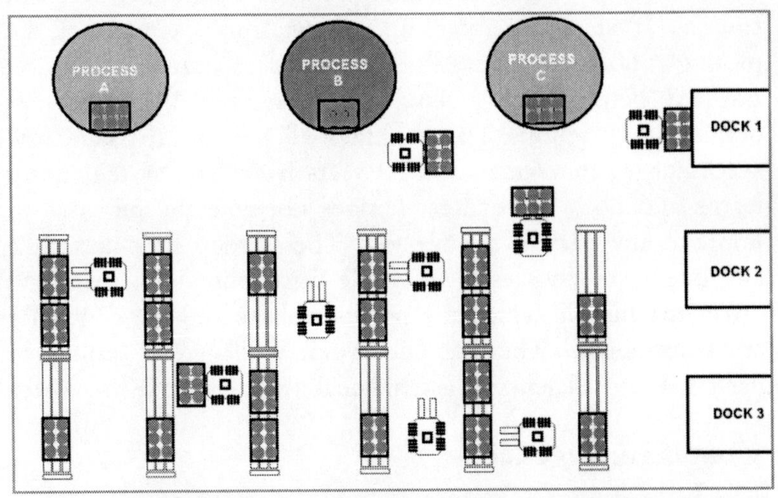

Example of the cheese warehouse automation

This example is just a tiny snippet of the complete specification of a control system, which is likely to be hundreds of pages and describes each facet of the project in great detail with drawings, requirements, outlines and milestones. When the project is finished we can observe how the robots work, and tick their operation off against the specification. Mission accomplished. The problem that needed to be solved, the solution and the path to the solution were all known at the beginning of the project, except for perhaps some minor variations that were made during the project's course.

An observer, who sees the many picking robots move around the warehouse, giving way to each other and picking the right cheeses to be processed, may get the impression that this is an intelligent system. Similar to a chess game, every action has been predetermined by a human programmer. The system will crash if it is faced with something that the programmer did not foresee. If an animal were to wander into the path of the picking robots it will very likely be killed. A human would not fare any better. Each picking robot performs the same pre-defined set of actions. The machines will perform the same task according to the same rules 20 years later. The system never learns and does not evolve. It does not take the initiative to improve any part of the process. The system does not make any intelligent choices. The whole warehouse is a cause and effect machine in which the picking robots and the controller are components. The machine is not intelligent. Learning in itself is not intelligence; it is what enables intelligence to evolve.

What is Intelligence?

What can we expect from an intelligent system, and above all, what is the definition of "Intelligent"?

Everyone you ask will have a different opinion about intelligence, and researchers are no exception. How can we begin to create a specification of intelligent systems if there is no consensus on the meaning of the word? It is necessary to return to the basic meaning of the word: The dictionary definition of "Intelligence" is given as the ability to comprehend, adapt, and process new information, to learn from experience and to apply that knowledge in a creative way. Two dictionary definitions of 'smart' are 'a sharp pain', or 'a quick mental ability'.

We all have seen the terms 'Intelligent' and 'Smart' applied to technology. We have 'smart phones' and 'smart' credit cards. In a TV commercial a car that senses road conditions is claimed to be "intelligent". But these devices work by the same principles as the picking robots that we discussed in our earlier example.

How about a chess or Jeopardy playing computer? Chess has a finite number of legal moves, and a defined number of fields. The program can calculate its next move, weighing off the value of pieces that it loses or takes and determines a strategy by calculating a number of possible moves and anticipated counter-moves. It also has a database of master games that can be searched for moves. It plays the game by computation, not by careful thought and skill. The IBM 'Watson' Jeopardy playing computer[14] impressed many people, but its creators are quick to admit that the machine is not intelligent. Speech recognition software interprets the question and converts it into several digital key phrases. Those digital 'keys' are then used to search a database. When the most likely answer is selected, it closes a contact to sound the buzzer and the answer is converted from digital text into speech. It is an extraordinary search engine and an impressive feat of engineering, but it is not 'intelligence'. These devices are not intelligent by the dictionary definition. They are machines that follow the rigid list of step-by-step instructions that have been defined by a human programmer. These devices are controllers, and no

[14] FEBRUARY 2011 © Stephen Baker. McKinsey Quarterly "The programmer's dilemma: Building a Jeopardy! Champion".

controller so far can claim to comprehend, adapt, or learn new information.

The confusion about intelligence has crept into academia; a paper on a Stanford University web site deals with the question "What is Artificial Intelligence"[15]. It states that intelligence is the "Computational part of the ability to achieve goals in the world". A thermostat is a simple goal seeking system: It maintains a more or less constant temperature around a set point and it has sensory feedback. Goal seeking systems are not intelligent. They perform a function that searches to reach target data,

A specification for an intelligent technology would need to include:

- A method of learning from experience and feedback at different levels.
- An evolving relational knowledge base. Relational means that the items in the knowledge base are linked, that a relationship between them exists.
- The ability to determine what is new and what is known from previous encounters or actions.
- The integration of existing, stored knowledge with newly learned knowledge.
- A way to communicate with the outside world, this could take several forms; such as a pulse train that is injected into spinal nerves, a connection to a computer, a filter-based speech synthesis chip, etcetera, depending on the application of the technology.

---

[15] http://www-formal.stanford.edu/jmc/whatisai/node1.html

- Stored 'knowledge' that is instantly accessible and without the need of a 'sorting algorithm' or sequential processing.
- Sufficient storage to store a 'lifetime' of experiences and knowledge. Alternatively the creation of new storage locations as they are required. The brain forms new storage locations by forming new synapses.
- A knowledgebase containing definable linked lists; one leads to another, and another, etcetera.
- Prediction. The next memory is recalled an instant before it is confirmed by sensory input. We experience this when we listen to music. We 'know' what the next note is supposed to be even though we may know nothing about music.
- Association between the senses, linking knowledge that originates from different sensory sources; such as a word sound with an object. For instance the sound 'book' with the thing we see and feel in our hand.

The Development of Human Intelligence

How do we acquire the amazing abilities of our adult brain? A human baby is born helpless, more helpless than most other mammals. A newborn baby has a relatively large brain, about the size of a gorilla's brain, but it cannot speak, solve any problems or create a work of art. What a baby does best, besides the more obvious physical stuff, is learn. It learns from all its sensory inputs. A baby needs to sleep a lot to process all this new information and to make it a permanent part of its knowledgebase. When we examine the brain tissue of a newborn baby under a microscope we see massive numbers of neurons, but relatively few connections. Connections, and new synapses, develop when we learn. The connections in the brain are much more than just links. Each connection is a memory location, which stores a value that is instantly recalled when an input pulse occurs. The whole brain is constantly buzzing with such pulses. Pre-programming through DNA defines only the most rudimentary functions by the time a child is born. Learning is the process by which we develop our brain and create a model of the world, in minute detail, stored in trillions of synapses. Exteroceptive senses are the 'external' senses that inform us about the world: eyes, ears, nose, taste buds and skin. Pain is an interoceptive sense, produced within the body. A third sensing mechanism is called proprioception, and it informs us about balance and where our limbs are in relation to our body. A baby's brain receives a steady input stream from all these senses. A young infant moves its limbs at random. It has no control over those limbs, but every time its limbs move, the brain gets feedback from the muscles about the position of those limbs. Through this feedback the brain builds a model of movements and limb positioning. It does not yet control the limbs; it learns at a subconscious level and refines a set of synaptic values that resulted in each feedback position through a process called 'Synaptic Time Dependent Plasticity' or STDP. Feedback strengthens or weakens synaptic values. It creates a set of values that represent each limb position through repetition.

Therefore brain is not a control system like a robot control circuit is, but a rather intricate information carrier and retrieval system. The brain learns cognitive tasks and motor control skills through action and feedback. There is feedback at multiple cellular levels and also through the environment and our senses. During our entire lifetime our brain adds new information within these frameworks, or 'processing centres' that were created early in life. The process of synaptic modification is known as neural plasticity. The stored information is instantly recalled and updated every time it is references, like when we catch a ball or undertake any other activity. The brain recalls information by 'association'; referencing input patterns with stored patterns. Usually during the first year of life this control system gets sufficiently sophisticated for the infant to stand up, and then walk. Learning takes place in a massive parallel fashion. At the same time that the motor processing network is trained, the baby learns to recognize sound patterns in its mother's voice. Hearing starts to function somewhere between the 22nd and 26th week of pregnancy. Our 'mother's tongue' is the first language we hear, even before we are born, and we have a predisposition toward that language. We have learned its specific characteristics long before we could talk. A baby's auditory cortex is trained to understand a specific language, first individual sounds (frequencies), and then vowels and finally complete words and sentences. During early development a baby will repeat sounds like "ba-ba-ba-ba". This trains the speech centres of the brain. The baby also hears itself and interprets the sounds it is making in its speech-processing cortex. Its visual cortex is being trained to combine shapes into objects. Somewhere deep in the brain objects are linked; we learn to associate and give meaning to words, shapes, and objects.

The environment that we are born into has a large impact on the distribution of knowledge and the level of each 'intelligence type' that we develop. The mind literally grows by what it feeds on. While structural brain features have an effect on the effort that it takes to learn a specific task, it does not determine what we learn. Mathematicians for instance often exhibit enlarged sections of the parietal lobe. It is not clear if these specific brain features are present before learning occurs, or that they develop because of learning. Intelligence test results are largely determined by our background, and can be greatly influenced by the person who crafts the tests. Many I.Q. tests presume a western education. An I.Q. test devised for one cultural group will not be suitable for another cultural group because the distribution of intelligence types will be very different.

Intelligence Types

Psychologist Howard Gardner[16] has described eight intelligence types. In each of these intelligence types the 'smartness' is acquired as a result of previous experience, plus natural ability, which may be determined by the individual's brain structure. It may also depend on how well we were encouraged as a child to participate in certain activities, our belief systems (what we have been told), and our interests. Each individual will score in each intelligence type, but at different levels.

- **Linguistic intelligence** ("the gift of the gab")
  This is the ability to use language effectively, an acute awareness of the meaning of words and the ability to express yourself in a concise manner.

[16] Gardner, H. (1983/2003). *Frames of mind. The theory of multiple intelligences.* New York: BasicBooks.

- **Logical-mathematical intelligence** ("number and reasoning ability"). The ability to analyse a problem, reason deductively and think logically. The ability to manipulate numbers effectively and the abstract relationship between numbers and symbols.

- **Visual-Spatial intelligence** ("visual ability")
  The ability to recognize colors, depth and shape, and to think in three dimensions, even though only part of the picture is actually visible. The image below the chapter heading is a typical example of Visual-Spatial tests.

- **Bodily-Kinaesthetic intelligence** ("body smart")
  The ability to move the body or limbs with a high degree of fine motor control, and to use objects or tools skilfully.
- **Musical intelligence** ("musical ability")
  Skill in the composition, performance and appreciation of music.The ability to identify rhythm, patterns and to read between the lines.
- **Interpersonal intelligence** ("good people skills")
  The ability to understand the desires, feelings, and the intentions of other people.

- **Intrapersonal intelligence** ("self awareness")
  Self knowledge, understanding ourselves, our feelings and who we are in the world. To use this knowledge to order our lives. Includes deep thinking, analysis, and pursuing goals.

- **Naturalist intelligence**("green thumb")

How we relate to our surroundings. The ability to interpret, understand and act on occurrences in nature, natural surroundings and interaction with animals.

Other research classifies as many as twelve intelligence types, recognising also emotional, sensory, intuitive, and creative intelligence. It is not clear how these human intelligence types relate to artificial intelligence. Computer-based Artificial Intelligence does not exist. What is currently called 'Artificial Intelligence' has no correlation with human intelligence. In synthetic brain based learning systems, intelligence develops to a level that is determined by the number of available synapses and processing nodes. Only a learning system that approaches the complexity of the human brain will be capable of evolving to this higher level. When this occurs it is likely to lead to a new branch of psychology dealing with machine psycho-analysis.

Belief Systems

Our behaviour and natural abilities are strongly influenced by our belief system. Our interests are shaped by the things at which we excel. If we believe, or are repeatedly told that we are 'good at' something, then sooner or later we will be. At this point some people may jump up and point out that highly gifted individuals have enlarged lobes in specific parts of their brain. Given that brain plasticity enhances those parts, this is not surprising; In Einstein the parietal lobe was enlarged. Did this give rise to his mathematical abilities or does his interest in this field cause the parietal lobe to be enlarged? Most likely there was a natural ability to start with, and further interest in the field caused this ability, and the associated brain lobe, to be further developed, just like a muscle gets bigger with exercise.

Strong belief systems can help us to excel or can hold us back from reaching our full potential. Be careful parents what you tell your children, or how you respond to a report card. It is likely to determine their future.

Brain Structure and Functioning

The brain contains our mind. The mind can be seen as the brain's 'software'; it consists of all our recorded experiences, feelings, and movements; the sum of everything that makes us an individual. This information is contained in synapses in formatted brain 'processors' and the structures that link it together. These structures look like trees with elaborate branches and root systems. If any part of the information tree gets triggered we recall the entire event. A familiar smell may trigger a childhood memory – good or bad. Often we don't remember something until we get a little hint. The moment we get that hint, we recall the entire story or experience. Hints can be in the shape of images, smells, tastes, words or sounds. We also usethis tree structure to predict what comes next – walking down the stairs our foot seems to 'know' where the next step is. Similarly, we immediately recognize a melody by a few notes. Our brain has already predicted the rest of the notes that we expected to hear. Hearing the note played out of tune causes an emotional response, something similar to pain. This 'pain' response is one of the mechanisms that drive learning.

Whether we are disassembling a motorbike or listening to music, our brains are constantly storing information that forms part of a memory tree. Memory trees are linked to form larger structures.

The same process works throughout our brain. Much of what we think we see is already present in our brain. We 'think' what we see. This is why eyewitnesses are so unreliable. When a group of people have witnessed a crime, investigators may get as many different stories as there are witnesses. There are numerous articles and books written about the differences in eyewitness testimonies depending on gender, age, race and perception. We see what we expect to see, not what is actually happening. We expect the event sequence to conform to the information already stored in our formatted memory tree. Our brains have a natural tendency to make things fit into existing memory trees.

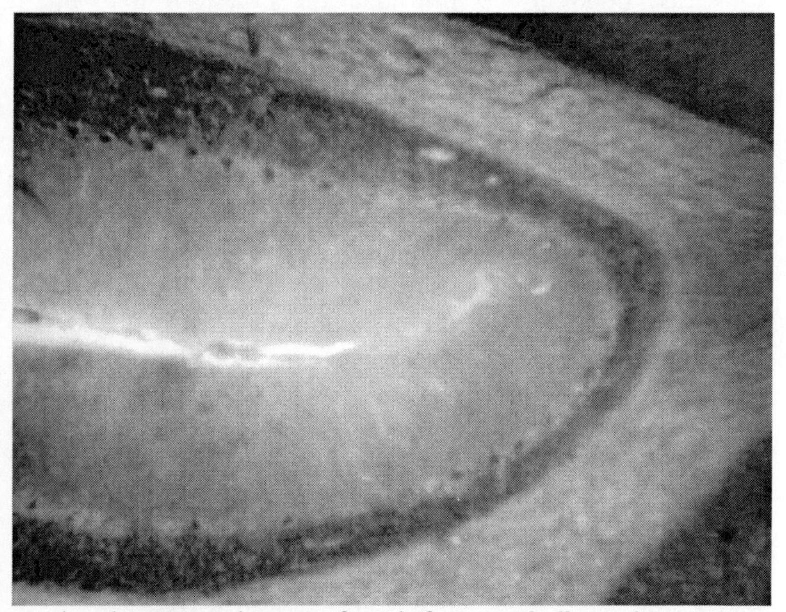

A microscope image of an infant cerebellum shows the layer structure of neurons, but not many connections. Image © 2012 by the author

Because we have a complex language, and the ability to convert visual symbols into words (reading) our mind works differently from other creatures on this planet. We think in words, our brain generates thoughts and speaks to itself. We underestimate how important language is in our thinking process. Only humans have developed a complex speech pattern. We use writing, a symbolic expression of speech, to preserve and transmit thoughts or experiences to others. Because of the hierarchical structure of the brain, language has a major impact on how we think. Try it; imagine that you cannot think in words. Blank all words out of your mind and try to think in feelings, pictures, and impressions without the use of words. This is nearly impossible to do because verbal expression is so near the base of the pyramid structure, a skill that is acquired early in life. Language has a direct impact on the formation of intelligence. Not that this makes one language group less intelligent than another group, just different. Language is interwoven with culture. This is visible in countries that have two or more languages, like Belgium. The French speaking part has a different feel to it than the Flemish speaking part. Writing is a symbolic expression of language. Written texts pass knowledge between individuals who live, or have lived, on different continents and thousands of years apart. Writing enables the transfer of knowledge across generations. Without language and writing tools we would still be using stone axes. To evolve our current level of intelligence we learned at many different levels. A child learns much before it is ready to go to school. Most importantly, a child will have language by this time, on which all other academic knowledge rests. Each generation can learn from the acquired knowledge of all the previous generations. We communicate abstract thoughts and mathematics through words. Language is what ties together what we observe with meaning.

Animal Intelligence

In cartoons and movies we 'humanize' animals with human speech and human behaviors. However, animals do not have such complex language or its symbolic representation to transfer knowledge. They think in feelings, impressions, and memories of events and act largely on instincts. They have a very limited set of noises that indicate danger, warn off competitors and enforce social status etcetera, which are similar to our non-verbal communications of grunts and nods. Anyone who has had a pet animal knows that they exhibit a certain level of intelligence. People who work with animals are generally able to interpret if the animal is in pain, aggressive, afraid or happy. This becomes more difficult to ascertain if the animal is not a mammal. At the most rudimentary level, consider the amoeba. It appears to have rudimentary memory[17] that stops the organism from going around in circles. There must be some means by which the amoeba determines whether to turn left or right, so it must remember which way it turned last. Though this is a single cell, it has complex mechanisms for food absorption, locomotion and reproduction.

Species	Average Brain Weight (grams)	Encephalization Quotient
Human	1400	7.6
Newborn Baby Human	375	
Dolphin	1550	5.3
Orca (killer whale)	5600	2.9
Elephant	5500	2.2
Cat	30	1.0

[17] "Smart amoebas reveal origins of primitive intelligence" New Scientist. 29 October 2008 by Colin Barras

Dog	72	1.2
Sperm whale	7800 ~ 9200	0.3
Chimpanzee	420	2.3
Horse	532	0.86
Mouse	0.63	0.50
Sheep	140	0.81

Table compiled by the author from a number of sources

Contrary to popular belief, humans do not possess the largest brains of any creature on earth. Animals like the Sperm Whale and the Elephant have much larger brains. The largest brain of any species was 9.2 Kilograms belonging to a sperm whale. An average sperm whale brain is about five times the size of a human brain, but a whale does not have anywhere near the same intellectual abilities as a human. An elephant brain is more than three times larger. It has been assumed that much of the larger brain is used in sensory processing because of their larger body size. A dolphin body to brain size is roughly similar to a human. Brain size is obviously not related to intelligence. Body to brain size is a factor, as is the quantity of grey matter (nerve cells) to white matter (myelin). The Encephalization Quotient or EQ method multiplies the body size to brain size ratio with a constant that is different for each species. Some small rodents have larger brains compared to body size. They would appear to be more intelligent than humans if body to brain size alone were taken into consideration. There is much debate about EQ calculations and how they should be applied.

Attempts have been made to teach apes sign language. They do not have the equivalent of human vocal cords, so that they cannot learn human speech. While they can learn to apply the signs for about a 1000 words, the hand gestures are sloppy, but they clearly understand and respond to the trainer's hand gestures.

African grey parrots have tiny brains but have amazing abilities in understanding and producing human speech, to the point where a trainer can carry on a conversation with the bird. Besides having remarkable speech abilities, there are parrots that solve puzzles, know colors and shapes, and can solve simple problems. Alex was such a vocal African grey parrot and his videos can be viewed on the Internet. I had the pleasure of briefly meeting with Dr. Irene Pepperberg[18] at a conference, and she spoke warmly about Alex, as if he was a much-loved child. Dr. Pepperberg has spent more than 30 years researching animal intelligence. Parrot brains have particularly well developed areas for speech and communications.

---

[18] Harper © 2008. Alex & Me: How a Scientist and a Parrot Uncovered a Hidden World of Animal Intelligence — and Formed a Deep Bond in the Process, by Dr Irene Pepperberg ISBN 978-0061672477

Chapter 6

The Function of Sleep

"There is a time for many words, and there is also a time for sleep"

Homer, The Odyssey

'Sleeping Robot' by David Gibson

Will a machine that works like the brain have to sleep, and happens during this time? To answer these questions we need to examine what sleep is.

Sleep is not simply a period during which the body and brain rest, as was once thought. The brain is very busy during sleep, as busy as when we are awake. Sleep is not the absence of brain activity but a critical psychological function essential for survival. Prolonged sleep deprivation leads to fatigue, depression, hallucinations, psychosis and eventually, death. Sleep deprivation does not only affect the brain and the cognitive functions, it also has an adverse effect on the body. The immune system is depressed, and there is an increased risk of diabetes type II, heart disease and high blood pressure. A person affected by sleep deprivation is more irritable, and has difficultly learning and retaining information. Specific parts of the brain, such as the thalamus and the pre-frontal cortex (perception) show decreased function during sleep, but the hypothalamus (metabolism) and pituitary gland (hormone balance) show increased activity.

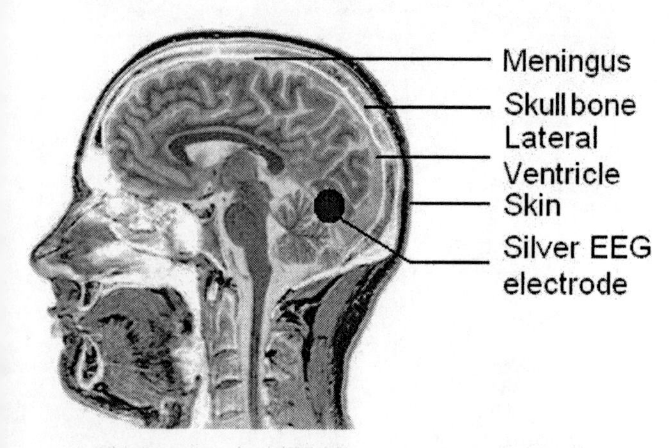

Obstacles to EEG brainwave detection

Stages of Sleep

Since the invention of the Electro-Encephalograph machine in 1927 it has been known that the brain emits low frequency waves, and that these waves are different during sleep from the waves that are observed when a person is awake. Measured through the lateral ventricle, the meningeal membranes, a thick layer of bone, and skin these waves represent an average of the output of millions of neurons. The wave patterns follow a distinct sequence of repetitive cycles of REM and non-REM sleep. REM sleep is distinguished by Rapid Eye Movements, from which it derives its name. The first 2-5 minutes of sleep is a transition period, sliding into light Non-REM sleep. This next stage consists of synchronized Alpha waves, followed by a short period of slow, high amplitude waves. The next stage of sleep lasts for about 20 minutes, with rapid rhythmic waves known as "sleep spindles". During this time the body temperature decreases and the heart rate slows down. This is followed by 30 minutes of deep sleep. This Non-REM sleep period is followed by REM sleep during which the voluntary muscles are paralyzed.

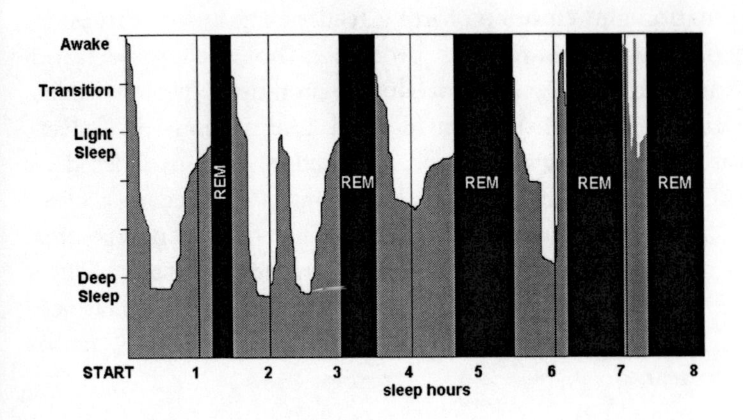

The motor cortex appears to be 'disconnected'. Vivid, colorful dreams occur during REM sleep. Brain wave activity is increased, as is the breathing rate. Average human sleep consists of 72% light sleep, 8% deep sleep, and 20% REM sleep. Babies sleep exists for 80% of REM sleep. REM sleep periods occur more frequently during the later part of sleep. The amount of REM sleep is increased followed periods of wakefulness with intense learning. There are many theories why we dream.

To consolidate memory, the brain has to generate its own stimuli. In its partitioned state, memories are played back. These signals cause stimulation of the limbic system, and the brain tries to interpret this internal activity assigning meaning to these signals. This results in subjective interpretation of signals that are generated by the brain. The information that makes up dreams is already present in the brain, and is not necessarily something that was learned the previous day. Increased activity is seen in the hippocampal area of the brain following intense learning. It is likely that the previous day experiences form the stimuli that recall existing, stored information. It is also likely that our desires influence the course of a dream, that stored information and emotions form a feedback path that drives the direction of a dream. This process is not well understood. During REM sleep, procedural memory (motor skills, conscious movement) is consolidated. During non-REM sleep, declarative memory (facts and knowledge) is consolidated. A physical clean-up takes place. The brain is encased in a clear, salty liquid, the cerebrospinal fluid. This fluid contains urea, dead neurons, and spurious neurotransmitters. Waste products are removed from cerebrospinal fluid by glial cells. Glial cells also remove incomplete formed synaptic connections (memories).

Wave Frequency	Name	Occurs during
0.5-4 Hz	Delta	Deep dreamless sleep
4 – 7.5 Hz	Theta	Light sleep, meditation, REM
7.5-14 Hz	Alpha	Deep relaxation, closed eyes
14-40 Hz	Beta	Conscious reasoning and thought
40 Hz up	Gamma	Insights, inspiration

No learning takes place during sleep. The brain areas that connect to the senses are inhibited and the brain is in an unconscious state. The brain appears to be partitioned with the connections between modules largely inhibited. Stimuli such as pain, bright light or loud noises are likely to awaken a sleeping person, which means that intense sensory information gets through.

The sleep requirements of animals vary and do not appear to correlate to intelligence. Insects do not sleep at all, although they may have rest periods. Bats, felines and rats sleep the longest, about 13 hours a day. Elephants, sheep and cows sleep 2-3 hours, while giraffes only sleep for a few minutes at the time.

A learning machine consisting of a synthetic neuro-anatomy of sufficient complexity will also need to have periods of memory consolidation, equivalent to what we call 'sleep'. The consolidation period is a function of the number of synapses in the synthetic brain. Simpler machines with few neurons, perhaps less than one million, may not need to consolidate memory, and thus, like insects, can go without a period of 'sleep'. The machine must have a provision for the general inhibition of large groups of neural circuits, and for the generation of stimuli to consolidate memory.

Chapter 7

Whole Brain Emulation

"Data is not information, information is not knowledge, knowledge is not understanding, understanding is not wisdom."

Clifford Stoll

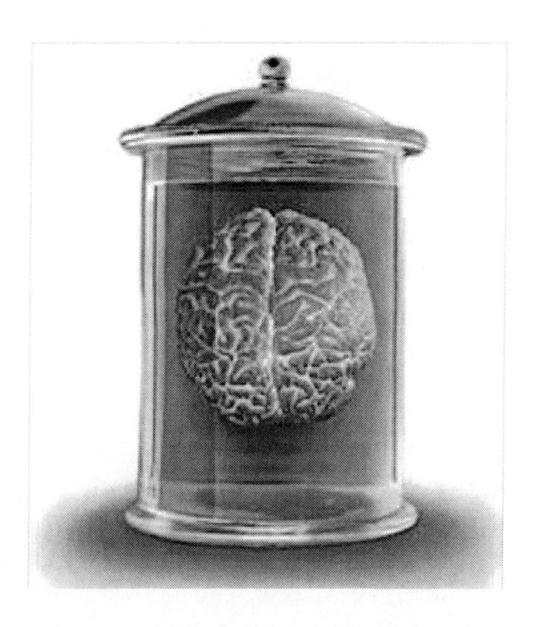

What is WBE

Its concept sounds too good to be true; scan the contents of an entire brain, copy it onto a hard disk and it can exist forever, though as a disembodied spirit in a computer. But is this possible, even at some time in the future, or is this science fiction? At this point in time we don't know how to build a brain, except by biological reproduction which involves very little or no science. Emulation means to imitate the behavior of one system or device on a different, and in most cases incompatible system. In contrast to simulation, emulation does not replicate just the output, but behaves in the same way and runs the same programs. Emulation is one step closer to the real thing than simulation. Think of this in terms of running an old game console program on a brand new laptop computer. The emulation program translates each original game console instruction into a processor instruction for the new PC. Communication commands and video graphics are translated and redirected to the appropriate devices. It is as if the game console exists as a phantom machine within a program on the laptop PC. In this way the emulator runs the entire program on a completely different computing platform. A simulator in contrast reproduces the same program behavior in a new program that has been created for the new PC. Whole brain emulation is a theory that assumes that the brain is some sort of computer (it's not). It aims to emulate the processes of the brain. This would involve uploading the knowledge (the mind) of a living person to an artificial platform that has an accurate representation of all the parts of the brain. To do this, the complete brain would need to be mapped including the entire connectome, all the neurons and all the information that is stored in the synapses. How can this be accomplished? Do we need to understand the entire brain and all its processes, including the mind, before we can upload it?

Copying the Unknown

Anders Sandberg[19] at the 2009 Singularity Summit discussed how a complex device can be copied without understanding it. He used the example of a computer chip. If we have the masks and the tools then we can copy the chip without a need to understand how it works. Even if we reverse engineer the chip by mechanically shaving layers of it and documenting its features we could make copies without knowing anything about its design or how it works. This is true, but I would question the validity of this example when it is applied to the brain. Copying the chip does not copy its programs or the data that is stored. We have a working intelligent system in the human brain that contains many terabytes of information. This information forms the conscious, sub-conscious and unconscious mind. A 'blank' brain does nothing, or very little. The brain of a newborn baby contains at least some information that has been placed there by DNA, and what has been learned in the womb. The brain contains 'hardware' in a network of neurons, synapses and supporting glia. Within that framework all our knowledge and our personality is stored, everything we know, from how to walk and talk to the mathematics and physics we learned in school. Every human brain is wired differently, not in a global sense but in detail. Even the brains of identical twins are not wired the same. In the case of a computer chip, we have exact details on how to apply the masks and every chip is wired exactly the same. The masks contain nice straight tracks of aluminum that interconnect the transistors to make the chip do what it is

---

[19] Dr. Anders Sandberg, Future of Humanity Institute, Oxford University

supposed to do. The brain is millions of times more complex than a computer chip, and its connections resembles a plate of very thin spaghetti. A neural cell has tens of thousands of inputs. Reverse engineering and reproduction works for microprocessor chips, mainly because even the most complex chips are super-simple compared to the brain and the software is easily stored and reproduced.

Scanning the Brain

To transfer all knowledge stored in the brain it must be scanned to create an exact map that includes the connectome and chemical composition of the synapses. This must be done on a living person without harm to that person, because the cells break down soon after death and the content of synapses is lost. A mathematical model of the brain would need to be contained in a computer that achieves more than 120 million times the performance of a current PC. This PC needs at least 100 Terabytes of memory and a program that emulates the whole brain model. Using an ultra-high resolution Magnetic Resonance Imaging scanner, we may be able to build such a whole brain model. The scanner must not only be capable of mapping cell structures and the many thousands of connections between them, but must also be capable to determine the level and chemical composition of neurotransmitters. This is relevant because knowledge is stored as levels of neurotransmitter chemicals. There have been advances in functional Magnetic Resonance Imaging (fMRI) that may make this scanning technology available in the future. An Ultra-high Tesla MRI scanner has a resolution of 50 nanometers[20] or better, but uses magnetic fields in excess of 15 Tesla. Such

---

[20] 1 nm = 1/1000,000th millimeter)

strong magnetic fields can have an adverse effect[21] on the brain by inducing currents in the iron-rich blood. This causes the blood to absorb energy in the shape of heat, similar to the way an induction cooker works. This would harm or kill the test subject and cause a detrimental distortion of the results. At this time there is no computer that has the capacity to emulate a complex model of an entire brain. The fastest computer in the world at this time has a performance of over 2.5 trillion calculations per second. Earlier attempts to emulate the brain have used the IBM Blue Gene supercomputer, capable of emulating about 5% of the human brain. The Chinese supercomputer is faster than the Blue Gene, so in theory it should be able to emulate 12% of the human brain, well short of the mark. Instead, a synthetic brain could be used in conjunction with a desktop PC to contain and operate a whole brain model.

These supercomputers are huge. The Tianhe 1A exists as 140 large fridge-sized enclosures and consumes megawatts of power. Multi-processing and multi-threading is an industry-wide trend that is also seen in our desktop and laptop computers. Microprocessor evolution has gone through several inclinations. In the early days it was driven by bus width. Processors went from 4 bit, to 8, 16, 32, and 64-bit busses. Then complexity was added to make processors faster, such as Internal cache, floating-point arithmetic, dual Arithmetic Logic Units, and predictive pre-fetch mechanism. Finally, in our current environment, the number of processor cores per chip is on the increase. Processors with 4, 8 or 16 cores are quite

---

21 "Biological Effects and Health Implications in Magnetic Resonance Imaging" Concepts in Magnetic Resonance, Vol. 125. 321-359, 2000. ALLAHYAR KANGARLU, 1 PIERRE-MARIE L. ROBITAILLE2. 1MRI Facility, 1630 Upham Drive, Columbus, OH 43210

common now. The Chinese Tianhe supercomputer contains over 7,000 NVDIA graphics processors in addition to 14,336 microprocessors. Each NVDIA graphics processor unit itself consists of 448 cores.

Given that the brain is a massive parallel processing system, with around eighty six billion simple processing cores, it appears as if computer technology and the structure of the brain are coming together. However, there are a few major differences to consider. The brain is not programmed by using an instruction list, nor does it have an address bus. The brain's data bus is millions of bits wide, while a computer's data bys is limited to 32, 64 or 128 bits. All knowledge structures are referenced in parallel through incoming data. The brain is a learning system that fills and modifies its knowledge structures that determine its processing algorithms. The brain consumes just 20 watts of power. It consists of much more than the cortex, or even the central nervous system. Sensory cells in the skin, the muscles and ears, eyes and the olfactory senses are all part of the nervous system. All our senses return massive amounts of information to the brain. The brain would not be able to function, let alone develop without these input streams. Emulation of the brain therefore needs to include the emulation of the senses and its feedback paths. A computer emulation of the brain cannot learn how to move an arm without spatial (proprioception) feedback from that arm. While much work has been done, there are still a large number of issues in relation to whole brain emulation. How much of the brain do we have to emulate? Some would say only the cortex, the outer layer of the brain where our awareness resides. Others believe that it is necessary to emulate the entire brain, because modules have a high degree of interdependency.

Chapter 8

Reverse Engineering the Human Brain

"If you invent a breakthrough in Artificial Intelligence so machines can learn, that is worth ten Microsofts."

Bill Gates, 2004

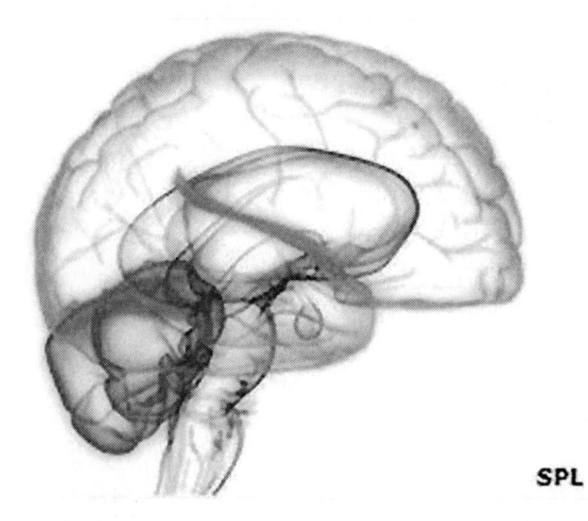

A 3D image of the brain, source: Surgical Planning Laboratory
© Harvard Neuroscience laboratory

The brain is the center of the nervous system. The nervous system branches out throughout the body with long nerves some of which are longer than a meter (3 feet). These nerves connect the brain to muscle tissue and sensory organs. They also carry feedback signals back to the brain. It is an information carrier that has a large and expanding capacity. The mind resides in the brain as electro-chemical values in trillions of synapses distributed throughout the brain. The mind has innate knowledge in minor DNA encoding, which was performed in the womb when the brain formed, plus everything we have learned since then. DNA just does not contain enough information to form the brain in all its complexity. The brain literally grows by what it feeds on, by what it learns. All our knowledge is stored in this huge structure of interacting, living cells. This is a volatile system; upon death the information carrier fails and the acquired knowledge disappears. Our brain consists of nothing but specialized cells. Everything the brain does must be explainable from neuron physiology, without being drawn into a computer 'processing' notion. This can be difficult to do in our present computer age, where we are used to think in terms of 'processing'. A computer is a logical machine and anyone who compares the operation of the brain to a computer is misinformed. The brain has no central processor, no separate memory module nor IO ports. The brain works by completely different principles. Computers are good at what they do, which is to compute, i.e. perform arithmetic operations on stored data under program control. Every letter and symbol has a fixed numerical index that is stored in memory at a determined, absolute address. A computer is never wrong, and it does not forget anything although a simple component failure can be catastrophic to stored data. Its programs may be wrong and produce the wrong results, but Boolean logic is infallible. A single faulty transistor out of millions will cause the entire system to fail. Using binary values, a computer uses an 'address' to point at each memory location to store or retrieve data. The brain has none of these concepts. The brain is floating in a clear salty liquid, the cerebrospinal fluid, which is produced in the brain at a rate of 500 ml per day. To control our body, the brain has to learn. Learning is a main function of the brain. It must not be overlooked in emulation systems. To build intelligent systems, it appears sensible to reverse engineer the only intelligent system that we are aware of, and then to build models of that system. A normal functioning human brain forgets, gets facts mixed up, comes up with the wrong answers and does not perform arithmetic very well or efficiently. Why would we want to copy that? Well, the brain also does some amazing things. It can create art, compose music, invent new solutions to problems and recognize people, places and objects in an instant. It can pick a face out of a crowd and an image from a busy background. It can control our hands to perform precise operations, to catch a ball, or to create a beautiful sculpture. It loses a few million neurons but the brain simply carries on as if nothing has happened. It can take initiative and learn and grow through new experience.

Overview

The human brain is a marvelous machine. In just 1.4 kg of wet, grayish-pink matter it performs tasks that no supercomputer in the world can match. We are only just starting to understand what processes are taking place in the brain, and how they lead to intelligent behavior. Assumptions about the 'control system of the brain', that have been made in the past are too simple, mostly due to a lack of understanding and suitable research tools.

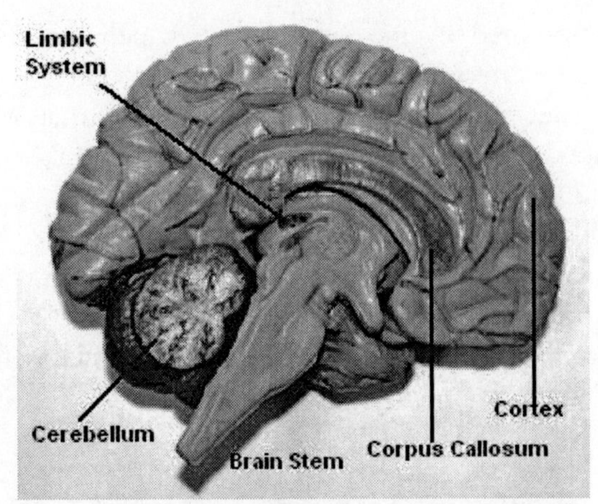

Sagittal view of the brain. Image by the Author

Most of its functionality develops from learning. The brain is not so much a control system as an information carrier. All its control functions are learned through multi-level feedback. The image of the brain shown here is a "Sagittal view", which means that it is sliced through the middle of the brain. The back of the brain is to the left. The large sphere of the brain is referred to as the cerebrum. The cerebrum consists of left and right hemispheres. The brain can be vertically divided into the inner limbic system, the outer cortex and the cerebellum, or 'little brain' and the brain stem. The cerebellum is located at the back of the brain, just above the neck, shown on the left behind the brain stem. The two hemispheres are connected through a large band consisting of approximately 250 million nerve fibers in the center of the brain, called the corpus callosum. In females this band is 10% larger, containing an additional 25 million fibers. The inner brain and the brain stem connect to the spinal cord, which is a large bundle of nerve fibers and neural cells that connect to the rest of the body and the organs. Each module has specific functions and interacts with other modules in the brain. The function of a module is largely determined by its connections, which originate at the senses and connect either directly to a module, or indirectly through other pre-processing modules. The major structure and location of brain modules is determined by DNA.

The wrinkled outer layer, the cortex, is only a few millimeters thick Unfolded it would look like a giant computer chip measuring about 33 by 33 centimeters or 13 by 13 inches. It contains more processing nodes per square millimeter than a computer chip.

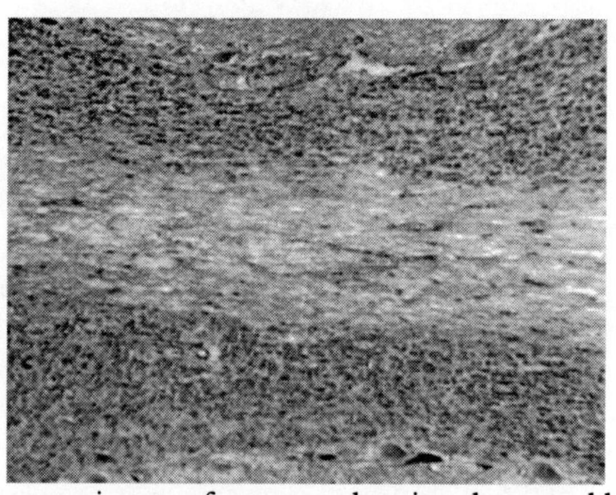

Microscope image of neurons showing the neural layers and cell types in the cerebellum.
Image by the Author

It is estimated that the entire brain consists of 85 to 100 billion neural cells that are connected through a 100 trillion synapses. Glial cells perform a support function to supply the neural cells with nutrients and remove dead neurons. The synapses form not only connections, but are also memory locations that store values. These values are recalled when an electrical signal, originating at a sensory cell, triggers the release of a stored neurotransmitter. The input pulse directly recalls the values that are stored in the synapses. This recall mechanism is time dependent with the value decreasing over time The neural cell integrates these time-dependent values. The connections from glia appear to synchronize the actions of groups of neurons.

The Cortex

Many Artificial Intelligence researchers are focusing on the cortex, the outer 'bark' of the brain. It is assumed that the folds and wrinkles of the cortex are caused by the creation of new connections between neurons that 'pull' sections together. These folds develop over time and are as unique as a fingerprint to each individual. Identical twins will develop unique folding patterns. The cortex is where all higher intellectual processes take place, such as conscious thought and conscious movement. Other regions in the cortex deal with auditory interpretation, language interpretation, visual association, conscious thought, and speech production. The cortex receives pre-processed information from the lower brain regions and passes conscious movement information to the lower motor regions of the brain. The cortex has six distinct layers with specific interconnects.

The cortex would simply not fit inside our skull if it were not folded. Inside this thin wrapping is a thick layer of myelin, which consists of proteins and fat. Large numbers of nerve fibers pass through this myelin layer to the lower brain regions. Myelin has insulating properties, much like the plastic on an electrical wire. Most nerve fibers are wrapped in a layer of myelin.

Limbic System

Beneath the myelin layer lies the limbic system. The limbic system consists of lower motor functions, emotions, long term memory formation and interfaces from the senses. It contains the amygdala, the hippocampus, the thalamus and other modules. The hippocampus performs a function that is crucial in the formation of long-term memory. It appears to cycle through the memories of the previous day to strengthen them and store them as long-term memory or to discard them. It is unlikely that these memories are stored in the hippocampus itself, because that would not be a function that fits in with neuron physiology. Knowledge must be contained in the cells that are connected, directly or indirectly, to the senses. The learning ability of a neural cell allows the synapses to be reused in the formation of new memories, causing earlier information to be forgotten. Long-term memory is formed when the learning function of synapses is suspended.

Brain Stem and Mid-brain

Inside the limbic system is the brain stem. The mid-brain is located on top of the brain stem and controls reflexes, limb movement and automatic functions, such as digestion and blood pressure. Nerves from the motor areas of the cortex pass through the brain stem to the rest of the body. The muscles of the face and neck are controlled by nerves in the brain stem. The brain stem includes the pons, the mesencephalon (midbrain) and the medulla oblongata. It forms a connection from the brain to the spinal cord. The pons transmits signals from the forebrain to the cerebellum. It also performs important functions in sleep. It 'disconnects' the spinal cord from the motor cortex, causing a temporary paralysis of limb muscles. Without this disconnection, a person may act out the actions that occur in a dream. REM (Rapid Eye Movement) is a sleep phase during which we dream. The pons signals the thalamus, which connects sensory information to the rest of the brain.

Cerebellum

The cerebellum is located at the lower back of the brain just above the neck. The cerebellum or 'small brain' is a coordination center for automated tasks and movement. The cerebellum has massive connections into the rest of the brain. Any tasksnot requiringany conscious thought is performed by the cerebellum. We do not recall these 'automatic' actions unless we force ourselves to be aware of them, in which case our cortex is involved in performing the same action.

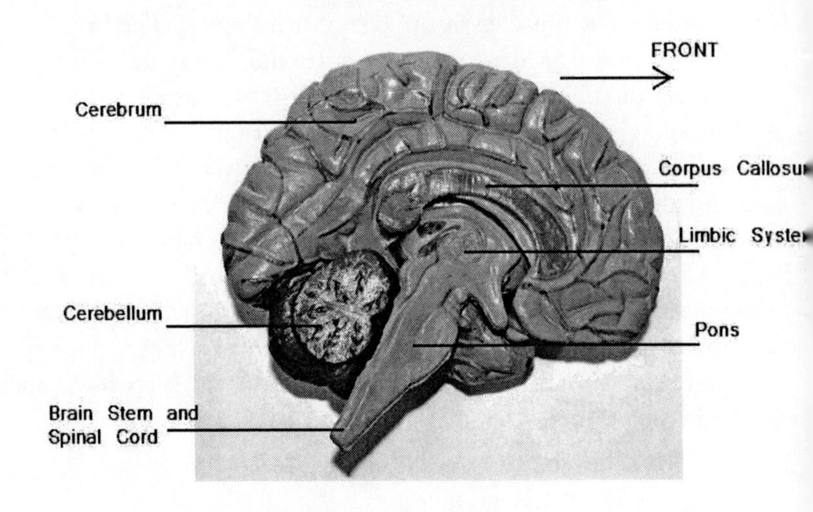

Sagittal view of the brain (cut through the middle)
Image by the author

Cerebrum

The cerebrum is the entire brain, except for the cerebellum. The cerebrum is covered by the wrinkly-looking cortex. Many of the functions located in one hemisphere are also duplicated in the other hemisphere, but the cortex has specific functional areas that have no equivalent on the other side of the brain. Interestingly, although women's brains are slightly smaller than men's, their corpus callosum (the band of nerves joining and allowing communication between the two hemispheres of the brain) contains an estimated 25 million connections, or 10%, more than in men.

The brain perceives the outside world through specialized synapses, found in the eyes, the auditory tract, the nose and tongue, the skin and in the muscles. On the output side, the brain connects to muscles through motor neurons. The brain's glial cells supply the neurons with nutrients, 'mop up' after a neuron has died, and create the fatty substance, myelin, that is used as an electrical insulation material around nerve fibers. Some glial cells have connections to neurons. Glial cells also facilitate the formation of new synapses. It is not clear what other functions glia perform in the processing method of the brain. The electrical connections between glia and neurons are thought to be a method of synchronizing neurons, much like the clock signal is used in a computer. Electrical connections exist between some adjacent neurons, and these also have a synchronizing effect.

Synapses

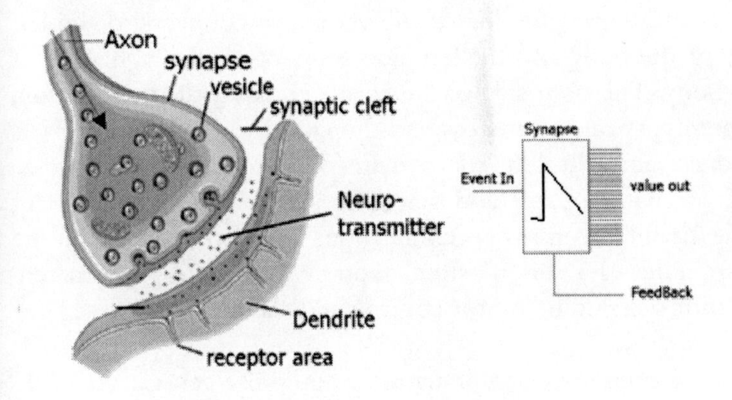

Image of a synapse and its symbolic equivalent image

The vast majority of connections in the brain are through chemical synapses, which are much more than just connections. A synapse forms a memory "value" that is triggered by the event pulse that originates at a previous active neuron. A single neuron is a processor node that can have as many as 200,000 synapses, or 'memory locations'. The average number of synapses is around 7,000. Synapses secrete a neurotransmitter into a gap, the synaptic cleft, when they are stimulated by an event pulse. The neuron's outer membrane has receptor areas, which receive the neurotransmitter chemical and cause a voltage difference in the cell's membrane. There are as many as 400 different neurotransmitters. Excitatory neurotransmitters such as AMPA (an amino-acid) cause an increase in membrane potential, whileinhibitory neurotransmitters such as GABA, cause a decrease. Fly sprays and nerve gas are neurotoxins that cause a disruption in the uptake of neurotransmitter chemicals in the synaptic cleft. The result is that the synaptic cleft gets flooded with neurotransmitters, and stops functioning.

The two large hemispheres covered by the cortex are called the cerebrum. The right side of the cerebrum connects to the left side of the body and the left side connects to the right side of the body. The right side of the cortex contains the imagination, creativity, spatial awareness and the emotional expression parts of the brain. The left side contains language, writing, logic as well as mathematical and scientific skills. The cerebellum or 'little brain' has massive connections into other parts of the brain and the brain stem, and coordinates 'automated' functions, including motor controland balance.

Brain neurons are signal integrators and pulse generators, while the gliaform a support and timing framework. Synapses form the connections between neurons and contain memory. To put this into perspective, imagine that you learn over a 30-year period. During that time you would have to 'program' 100,000 synapses every second to fill your memory capacity, were it not that the brain creates new synapses when they are needed. Since new synapses form all the time there is no real limit to the brain's memory capacity. Neurons combine synaptic data into a model. Initially input data is stored, when the whole brain formats itself and builds a 'model of the world' during early childhood. Subsequent input events trigger stored information in the synapses. Unknown input triggers an emotional response that updates the existing model. Synapses are attached in huge numbers to the dendrites of neurons. In a simplified model the dendrites may be thought of as the neural cell inputs. Each neural cell has one output called an Axon. The axon has many branches, each terminating in a synapse on many other neurons. The neuron looks somewhat similar to a bush, with a large root structure beneath and branches on top. The synapse is both a memory and the connection between neurons. Each synapse stores a chemical 'neurotransmitter' in vesicles, which is produced by the previous neuron and is released to the synaptic cleft on the next neuron. Glial (Greek for glue) cells were thought of as the framework that supports and nourishes the neurons. The human brain consists of nothing but specialized cells that integrate synaptic values and communicate with one another. There is no processor, no stored program and no separate memory module. Since the brain consists of nothing but nerve cells, everything it does has to be explainable from the neuron physiology. If we cannot explain its function from neuron physiology, then the doctrine is wrong, and we are missing something. The brain is the 'hardware' that executes the 'software', the mind. The brain's synapses store knowledge.

Brain Quick Facts

Average weight	Approximately 1.35 Kg
Average volume	1,350 ml
Size	14 cm x 16.7 cm x 9.3 cm
Number of neuron cells	Estimated 85-100 billion
Number of glial cells	Estimated at 1 trillion
Number of synapses	Estimated at 100 trillion
Neo-cortex neurons	Approximately 20 billion
Average neurons lost	One per second
Wave frequencies	Beta: 13-30 Hz. Alpha: 8-13 Hz Theta: 4-7 Hz. Delta: 0.5-4 Hz
Max number of synapses on one neural cell	~200,000 on Purkinje cell
Average number of synapses per cell	7,000
Nerve fibers in corpus callosum	250,000,000 (more in females)
Cerebrospinal fluid volume	150 ml
Production cerebrospinal fluid	500 ml/day
Cycling of cerebrospinal fluid	3 - 4 times a day
PH of cerebrospinal fluid	7.3 (Seawater = 7.2)
Optic nerve fibers	1.2 million
Brain weight vs. oxygen consumption	2% of body mass, 20% of total available oxygen consumption.
Fast signal transmission	430 Km/hour
Brain size: Human	2-3 times the size for the same

versus other primates	body dimension.
Variation in human brain size	.900 kg to 1.850 kg. Correlates to body size, no indication of intelligence.
Cerebrum	Large walnut shaped brain consisting of two concentric halves.
Cerebellum	'Small brain' at the back of the cerebrum
Limbic system	'Inner brain'
Brain stem	Connects to the spinal cord and regulates many automated functions.
Spinal cord	Body's 'information highway'that connects to the limbs and organs.

In recent years it has been found that glial cells also play an active role in the timing of neural transmissions and that many have synaptic connections to neural cells. Glial cells vary in connectivity, size and function. Neural cells vary in size, the number of synapses and the firing patterns that are produced when they are triggered. All these different cell types are organized in columns. A neural column learns to respond to a complex sensory pattern. The columns are structured in a hierarchy, forming a huge multi-layered content addressable memory in which stored knowledge determines the selection criteria for subsequent layers.

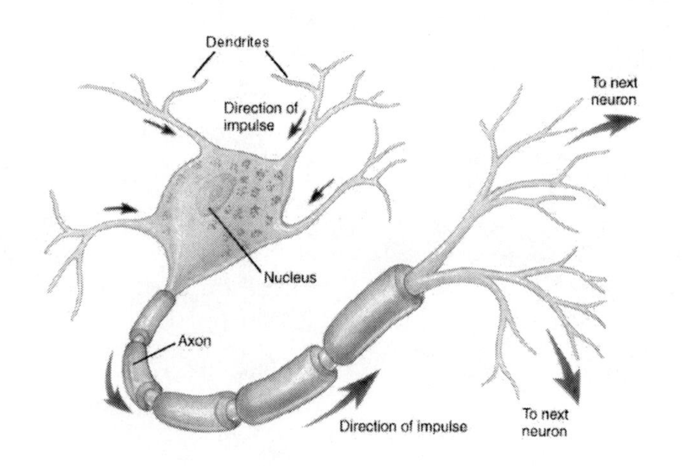

Diagram of a neural cell

The routing through this huge matrix depends on the knowledge that is stored in the synapses.

Let us consider for a moment what would happen if we were to simulate a complex brain model on a computer. The time for a 3 GHz PC to retrieve synaptic values, calculate their integrated value, simulate the neuron and determine the axon state with feedback to update the synaptic values is about 10 milliseconds, so we can calculate 100 synapses a second. A single pass, one-second brain simulation for a whole brain model with 100 trillion synapses would take 317 years to complete. This is an optimistic figure because the synapses are updated many times a second.

The brain model gets increasingly complex when we consider the more than 400 different chemical neurotransmitters and peptides. Different neurotransmitters cause dissimilar effects in the receiving neuron. There are inhibitory and excitatory neurotransmitters with some causing long-lasting and some very short-term effects. Glial cells differ in function; they regulate the myelination of axons, support neural cells, and contribute to the learning of training pattern values in synapses. Neural cell types differ depending on their location within the brain and their function. Much focus has been placed on the outside layer of the brain, the cortex, where intelligence and conscious thought are understood to reside.

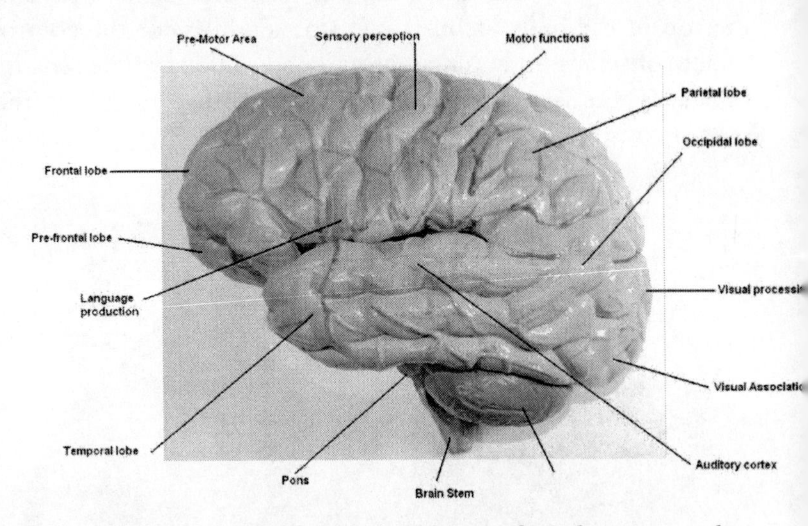

The cortex is the large wrinkly looking surface that covers the entire brain. Underneath the cortex the cerebellum, the brain stem and the top of the pons are visible

Directly behind the cortex lies the cerebellum. It consists of a dense population of neurons. Injury to this part of the brain results in the inability to judge distance and the loss of coordinated movement. People who have their cerebellum removed have difficulty balancing, are clumsy and suffer muscle spasms. The cerebellum is easily damaged by ethanol (common in alcoholic drinks). This type of damage is also found in babies that have been exposed to ethanol in the womb.

Underneath the thin layer of the cortex there is a substantial layer of myelin, a fatty substance that contains many nerve fibers that connect the parts of the brain and acts as an insulator for electrical signals passing through axons. The limbic system, or mid-brain, lies between the myelin layer and the top of the brain stem. It contains modules for the control of subconscious body functions and connections to the sensory organs. There is a second copy of each limbic module in the other hemisphere of the brain.

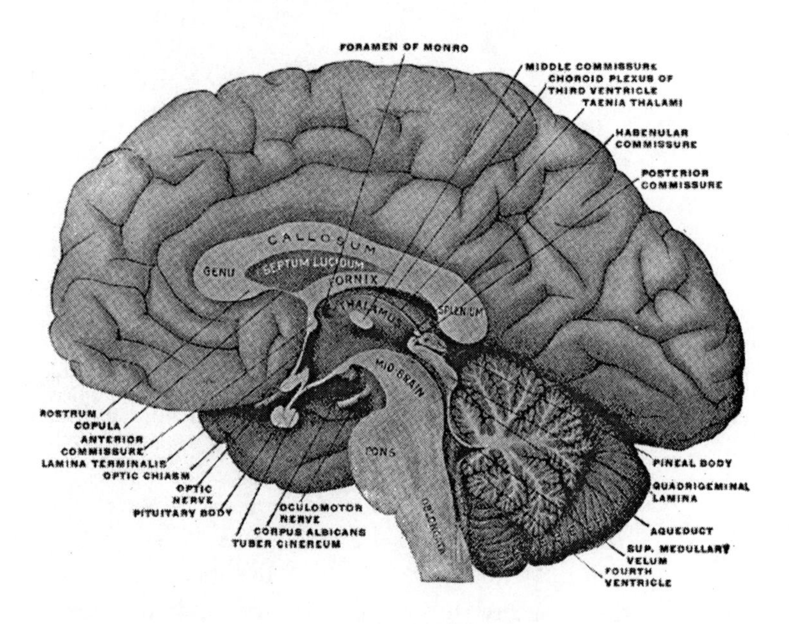

Limbic system module	Function
Thalamus and LGN	Sorting and distribution center for the senses. The LGN (lateral geniculate nucleus) connects to the optical nerve, the midbrain and V1.
Hippocampus	Conscious recollection of past events, indexing service. Damage to the hippocampus causes a condition in which a person cannot form (or recall) new long-term memory while existing memory is unaffected.
Amygdala	Processing and memory of emotional events. Aggression, fear, anger and sexual response.
Fornix	A band of nerve fibers that connects the hippocampus to the hypothalamus.
Cingulate gyrus	Sit on top of the corpus callosum. Autonomous functions for heart rate, blood pressure and cognitive processing.
Corpus callosum	A wide band of nerves that connect the two brain halves together. Separating the two brain halves by cutting this band causes a condition called split brain.
Pituitary gland	Secretes hormones into the

	bloodstream.
Hypothalamus	Links the brain to the glands that secrete hormones into the body. Controls hunger, appetite, thirst, response to pain, pleasure, anger, and sexual satisfaction. Responds to blood pressure, gut tightness, darkness, and blood temperature.
Basal ganglia	Links the cortex with the thalamus. Inhibitory function in motor control.

The limbic system is wrapped around the brain stem. The brain stem contains the midbrain, pons, and the medulla oblongata. The brain stem connects the cerebellum and the cerebral hemispheres to the spinal cord. In adults the spinal cord is 43-45 cm long, and contains an estimated one billion neurons. The spinal cord carries motor signals and sensory information to and from the body. Most sensory signals (the exception being smell) pass through the thalamus. The brain stem and spinal cord also control bodily functions, such as breathing, blood pressure, swallowing, and heart rate.

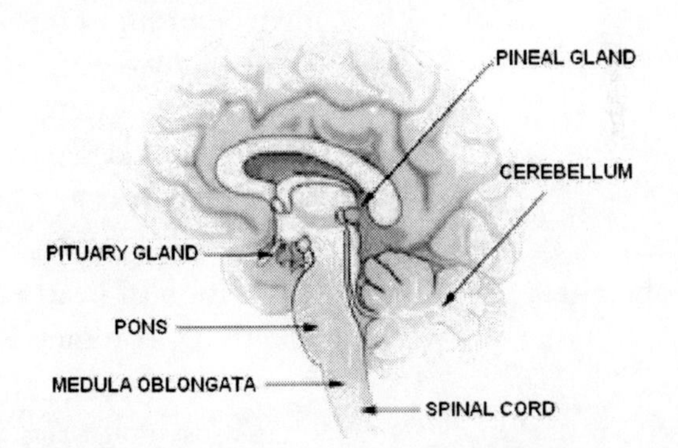

Our emotions are governed by the limbic system. The limbic system controls glands such as the sweat glands, the adrenal gland (near the kidneys), the salivary glands and the pituarygland, which secretes growth hormone and regulates metabolism. Sensory information is passed to the limbic system before it reaches the conscious cortex. This has some interesting effects. A person who has severe brain damage in the upper brain regions will still turn his or her eyes to a person walking around in the room. Our eyes are attracted to a moving object before we make a conscious decision to look. Our emotions are roused before we have a chance to think about them. The cortex receives input some time after the limbic system. Our body language is also dominated by the limbic system. We have little awareness of our body language; the cortex is unaware of our body language unless we make a conscious effort through awareness training.

Brain Stem module	Function
Midbrain	Focuses the eyes on moving objects. Hearing and body movement.
Pons	Relay station for signals from

	the cerebrum to the cerebellum. Sleep, bladder control, swallowing and respiration. Also taste, eye movement, posture and facial expression.
Medulla oblongata	Regulation of heart rate, breathing, vomiting and blood pressure.
Spinal cord	Transmission of signals from the brain to the body, some evidence of autonomous functions such as walking. There are 1 billion neurons in the spinal cord.

Massive numbers of connections exist between all of the brain's modules. Everything is connected. Not every brain module is constructed in the same way. The cortex has 6 layers.The structures of the limbic system, which areenclosed by the cortex, generally havefewer layers. The Hippocampus for instance has 5 layers.

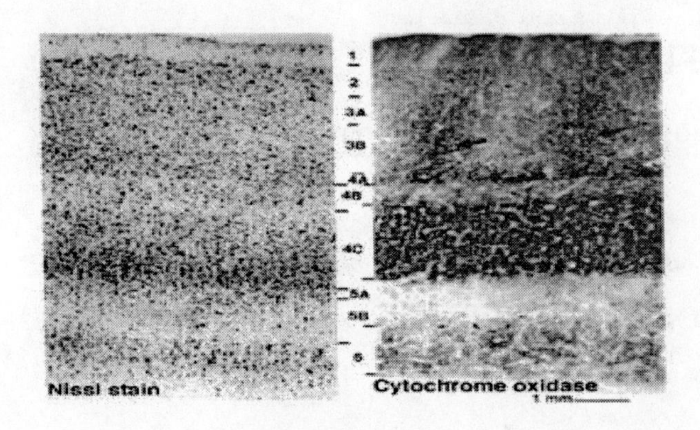

The layers are made visible by different staining techniques as shown here. The layers have no clear boundaries. During early infancy another layer of neurons exists in the 'white matter'. That layer disappears during early childhood. Sensory input is connected to the neurons in the dense middle layers, which are separated into seven sub-layers.

Chapter 9

Biological and Artificial Sensory Organs

"All our knowledge is the offspring of our perception."

LEONARDO DA VINCI, Thoughts on Art and Life

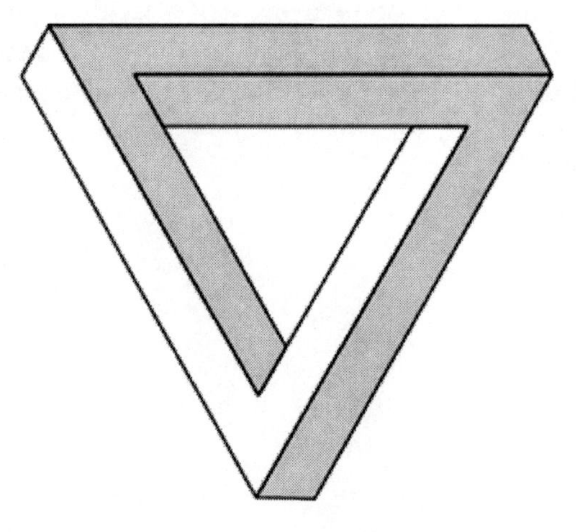

A 2D sensory illusion, the Penrose Triangle (public domain)

There is very little said about sensory perception in discussions about Artificial Intelligence. This is surprising since any artificial brain will require not just sensors, but biologically accurate sensors of sufficient complexity to allow the brain to function. To learn about the natural world, more information is required than what comes from a microphone or from a video camera. The organs that form our senses are complex and sophisticated and transmit information in the form of pulses. Sensory devices need to accurately emulate these natural senses, complete with their feedback paths. A sensory device can only train an artificial sensory cortex to develop a 'processing center' when a biologically correct sensing device is used. What will be the future of such artificial sensing devices? Could they perhaps be used in artificial brains as well as in prostheses for humans? We will examine each of the human senses to see what suitable artificial senses are available.

Introduction to sensory devices

Besides a much higher resolution and sensitivity than a video camera, the eye provides color, intensity and edge detection pulse streams. The tongue not only provides taste information but also information about food texture. The ear performs spectrum analysis, intensity, balance and acceleration functions. All this sensory information is communicated as bursts of electrical pulses. Even though we perceive light, no light enters the brain. The entire nervous system functions through the transmission of timed electrical pulses. There is no relation between computer image data and the visual information that is passed through the optic nerve, or audio data streams and the information provided by the cochlea. This information is much richer in context, providing more than just amplitude and frequency. Data is contextual and restrictive in that it can only express a limited amount of information. It is a symbolic expression, in ones and zeros, of information as illustrated by the following example.

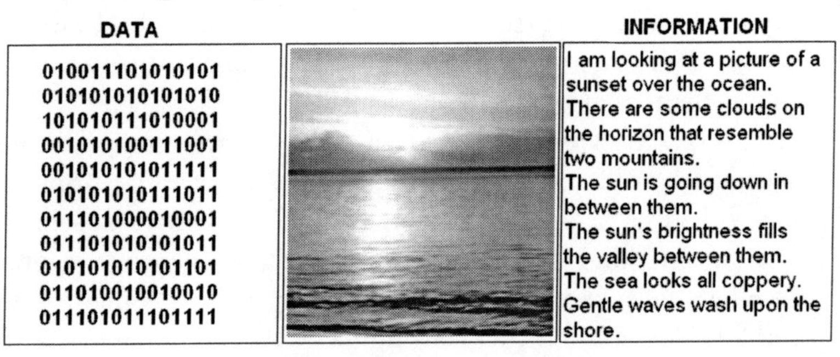

Data vs. an image vs. information.

The image shows a sunset. A human who perceives the sunset may remember it in words as it is described on the right. Given the verbal information we can form an image in our mind. We store both the verbal and the visual impression as analog values across many millions of synapses.

```
0000: 42 4D 36 43  00 00 00 00  00 00 36 04  00 00 28 00  BM6C......6.
0010: 00 00 7D 00  00 00 7E 00  00 00 01 00  08 00 00 00  ..}...~.....
0020: 00 00 00 3F  00 00 00 00  00 00 00 00  00 00 00 00  ...?........
0030: 00 00 00 00  00 00 00 00  00 00 00 00  80 00 00 80  ............
0040: 00 00 00 80  80 00 80 00  00 00 80 00  80 00 80 80  ............
0050: 00 00 C0 C0  C0 00 C0 DC  C0 00 F0 CA  A6 00 00 20  ............
0060: 40 00 00 20  60 00 00 20  80 00 00 20  A0 00 00 20  @.. `.. ...
0070: C0 00 00 20  E0 00 00 40  00 00 00 40  20 00 00 40  ... ...@...@
0080: 40 00 00 40  60 00 00 40  80 00 00 40  A0 00 00 40  @..@`..@...@
0090: C0 00 00 40  E0 00 00 60  00 00 00 60  20 00 00 60  ...@...`...`
00A0: 40 00 00 60  60 00 00 60  80 00 00 60  A0 00 00 60  @..``..`...`
00B0: C0 00 00 60  E0 00 00 80  00 00 00 80  20 00 00 80  ...`........
00C0: 40 00 00 80  60 00 00 80  80 00 00 80  A0 00 00 80  @...`.......
00D0: C0 00 00 80  E0 00 00 A0  00 00 00 A0  20 00 00 A0  ............
00E0: 40 00 00 A0  60 00 00 A0  80 00 00 A0  A0 00 00 A0  @...`.......
00F0: C0 00 00 A0  E0 00 00 C0  00 00 00 C0  20 00 00 C0  ............
0100: 40 00 00 C0  60 00 00 C0  80 00 00 C0  A0 00 00 C0  @...`.......
0110: C0 00 00 C0  E0 00 00 E0  00 00 00 E0  20 00 00 E0  ............
```

Part of the same image shown as a computer memory dump

If we close our eyes we can recall the image from memory and see the image in our 'mind's eye'. We don't see pixels (the computer picture elements of single colored dots, represented by 1 or 0) but objects like clouds, ocean, waves and sun. A computer stores the sunset as a compressed bitmap of pixels, a fraction of which is shown below and to the left. Extracting and comparing objects from a data set is a lengthy and difficult programming task that has nothing in common with the way that our brain performs this task.

Sensory Perception

The brain image recognition system receives edges, shapes and color streams. One of the functions of the eye is to perform edge detection, and these edges are combined into shapes in the visual cortex. The brain interpolates between missing line segments so that we recognize the whole object even if some of its shape is obscured. The same information set always recalls the same knowledge through association. Because our brain works with shapes we recognize objects no matter their size, distance or orientation. A shape is constructed from line segment information and color information. We recognize objects by their function as well as by their shape. Anything that has a platform at seating height, suspended on one or more legs, will be considered a chair, no matter how weird its shape is.

Five Senses?

Aristotle defined in 300 BC that we only have five senses; smell, taste, hearing, sight and touch. In fact, we have many more. We are aware of some of these, and not of others. We know where our limbs are in relation to our body, a sense called proprioception. We have a sense of acceleration and balance in our inner ear, the same mechanism that can cause us to get seasick. Besides touch, our skin senses pressure and the temperature in the surrounding air, and pain if we get injured. The passage of time is perceived through our inner biological clock, e.g. we sense the duration between events. Then there are internal senses, such as pulmonary stretch sense that gives the brain feedback about breathing. The brain receives information from pressure sensors in the kidney's renal artery to regulate our heart rate and maintain blood pressure, carbon dioxide sensors in the blood vessels to regulate breathing rate, and receptors in our bladder and rectum that inform our brain that its time to visit the toilet. In addition to these, some animals have senses that we don't have. Pigeons sense the magnetic field of the earth. Electroception, the ability to sense an electric field is found in sharks, which use it to find their prey. They can detect electric fields that are extremely weak; at 5 nV/cm ($5*10^{-9}$) this is the equivalent of sensing a 1.5V battery from a distance of 3,000 km. The electrical field is distorted by objects that conduct, while resistive objects spread the field. Bats use echolocation, a form of sonar, to 'see' a picture of sound vibrations. They are capable of detecting the differences in the echo that is returned from objects to determine where objects and their edges are in a dark space. An artificial brain could be equipped to see heat radiation, or any type of radiation in a spectrum that is invisible to humans. Whatever we want to use, we must follow the same 'rules' as the human or animal senses; each sense needs to provide feedback and pulsed information, rather than data.

How much of all this sensory apparatus is required to get an artificial brain to develop and interpret its environment? How well can we emulate them? Sensory perception is very important to brain development. As we have seen, the brain develops by learning. It constantly grows new synapses and connections, and stores knowledge. Learning requires either repetition or intensity of the stimulus, and massive amounts of feedback. All sensory perception is a source of learning. A brain without any sensory perception at all would do nothing, and would most likely not develop. It would be as useless as a microprocessor that is not connected to any peripheral or memory devices. The human brain receives input from a great number of senses, and feedback from every neuron, every column of neurons and also feedback through multiple senses; if we move our hand we see our hand move, while the brain also receives feedback from the muscles in our arm and hand. All this information trains the brain, and causes the brain to format itself into processing centers.

Visual Sense and the Eye

We don't think about the complexities that make up our senses, we largely take them for granted. The eye is an integral part of the central nervous system, which receives electromagnetic waves in the spectrum of 'visible light'. These waves are reflected by all objects around us. The eye turns them into pulse streams that the brain interprets in the form of pictures. The eye provides much more information about an image than a camera. A camera translates individual color dots into a serial stream containing intensity and color data. At around 16 Megapixels, they have about a seventh of the resolution of the eye (although the eye does not take static pictures). The eye moves rapidly in a motion called saccades. When we look at an image we move our eyes rapidly to focus on points of interest. The movement is not the same in all cognitive tasks. In face recognition, we focus on the eyes, nose and mouth. The center of the fovea at the back of the eye contains the largest number of photoreceptor neurons. The brain combines information from rapid eye movements with a resultant resolution in the vicinity of 600 Megapixels. The ganglia and bipolar cells in the retina perform edge detection. The cone cells are receptive to color information and the rod cells receive a black and white image. There are wide area receptors and narrow area receptors. The number of photoreceptors decreases exponentially from the fovea in the center of the retina to the edges. To get an artificial brain to learn to recognize shapes we will need to provide similar image information that the brain receives, and therefore use an artificial retina that is similar in function to a biological retina as image detector and pre-processor. Dots perceived by individual photoreceptors are strung together into lines. Higher in the hierarchy, these lines are combined into shapes and objects, and higher still they are combined with color. Everything in the brain is stored in such hierarchical arrangements. Things in the world happen in sequence; one step follows after another and one event gives rise to another. Our brain is aware of these sequences. We 'expect' one event to be followed by the next. Therefore the brain has a prediction mechanism. The brain has already predicted where the next step is when we walk down the stairs. It does this by pre-sequencing the next column of information in the hierarchical storage set. This is why optical illusions work. Our brain 'expects' to see something, and we are seeing what we expect to see rather than what is really there.

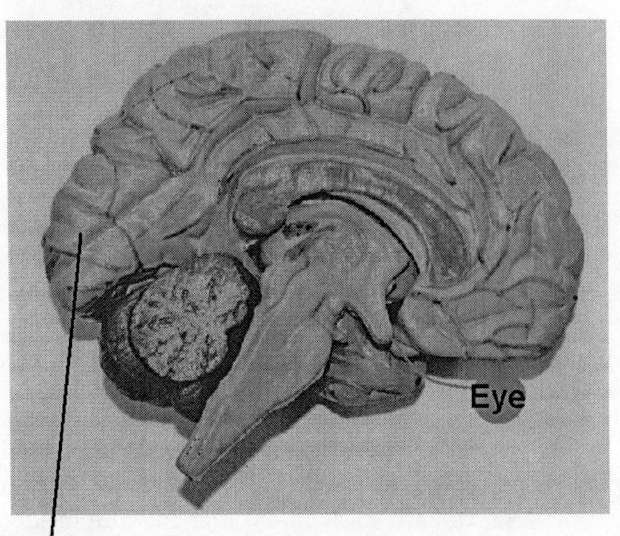

The location of the visual processing areas of the brain

Physiology of the Eye

A large part of the brain and the cortex is dedicated to visual processing. The visual cortex is located at the back of the brain and consists of the areas V1 to V4, MT and the Geniculate Nucleus. In the eye, light is first passed through the cornea, the pupil and the lens. The iris is the visible part of the eye that has color. It contains muscles that pull to enlarge or reduce the size of the pupil. The pupil changes size depending on the intensity of light that is received by the retina.

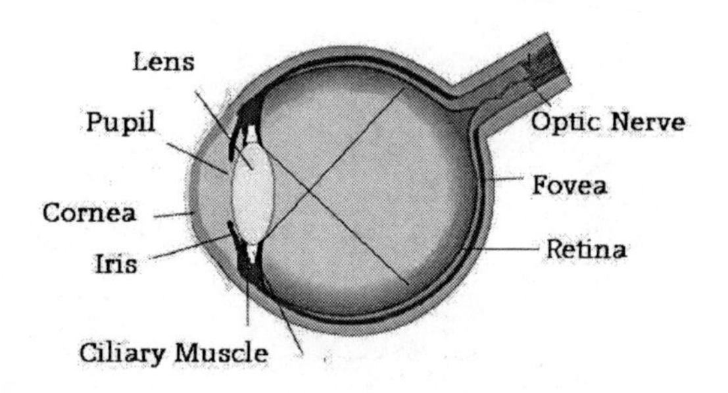

The human eye. An image is projected at the back of the eye, the retina.

The image is projected upside down on the retina, after passing through the lens and the clear liquid that fills the inside of the eye. The image on the retina is kept in focus through ciliary muscles that change the shape of the lens, which has the consistency of jelly. The retina has nine distinct layers and is considered to be part of the Central Nervous System. Light has to pass through eight layers before it reaches the photoreceptor neurons, which are in the bottom layer and pass their signals up to the top layer.The photoreceptor neurons come in two shapes;rods and cones. The rod cells outnumber cone cells by about 20 to 1. There are just over 90 million rod[22]cells in this layer, concentrated in the periphery of the retina. A rod cell responds to a single photon of light in the blue-green part of the light spectrum. Cones are densely packed in the fovea, the center of the retina. There are approximately 4.5 million cone

---

[22] Osterberg (1935). "Topography of the layer of rods and cones in the human retina," Acta Ophthalmol., Suppl. 13:6, pp. 1–102.

cells in the human eye[23]. They are 100x less sensitive to light, and are used in color vision. Each cone cell responds to green, blue or red light. The eye does not only receive image information, the processing of visual information begins here. The eye contains neural cells that detect a change from light to dark, e.g. edge detection.

Artificial Retinas

Several projects are under way to produce an artificial retina and artificial eyes. Some patients have had test devices implanted that restore very limited vision. Present-day devices are primitive. It is hoped that in the future the entire prosthesis, including the camera, can be implanted in the eye. The signal from a camera, mounted on a pair of glasses, is processed and sent wirelessly to a receiver in the patients' eye.

Image of the Nano-Retina CSEM bionic chip implanted in a patient's retina, showing how the photoreceptor cells arestimulated with electrical pulses.

---

23 1999 C.W. Oyster "The human eye: structure and function". Sinauer Associates

A thin-film silicon chip is attached to the patients' retina so that it forms an electrical bond with the neurons. The chip contains a receiver and contact pads that provide electrical stimulation. The patient's retina must contain functioning neurons to use one of these devices. While they restore sight to previously blind people, and better resolutions are expected in the near future, these devices clearly do not replicate the function of a biological retina. They stimulate the neurons that are still there buthave lost their photoreceptor properties. More relevant research is focusing on a device that will produce a signal that can be inserted in the optic nerve[24]. The 'Brains in Silicon' group at Stanford University has developed a replication of the function of the retina, and produced a signal that could be injected into the optic nerve. This research could lead to an implantable retina in the future, and is useful to provide vision, and equally important, sufficient information to train an artificial brain.

Olfactory Sense: The Senses of Smell and Taste

The sense of smell is far more important than we realize. Our food loses much of its taste when our nasal passages are blocked because of a cold or flu; we smell our food as much as we taste it. Foranimals in the wild, it is a crucial survival and hunting tool. Humans also possess the ability to sense danger by smelling, but we are largely unaware of it. People exposed to the smell of spiced apple for instance experience a drop in heart rate and blood pressure. Likewise, the smell of strawberries has a calming effect. People with darker skin are

---

[24] K A Zaghloul and K Boahen, A silicon retina that reproduces signals in the optic nerve, Journal of Neural Engineering, vol 3, no 4, pp 257-267, December 2006.

more sensitive to smell. Air that contains molecules of substances is sucked into the nasal cavity and passes the nasal epithelium. The yellowish-brown epithelium contains about 10 million receptor cells, olfactory receptors that are sensitive to chemical stimulation. Each olfactory receptor is a bipolar neuron that has hair-like dendrites that protrude into the mucus covering the epithelium. An average human can distinguish between 4,000 different smells. Trained noses can distinguish up to 10,000 different smells. Each smell is made up of different chemical compounds. Molecules bind themselves to individual receptors, which signal the olfactory bulb in the limbic part of the brain.

Artificial Noses

Bomb and drug-sniffing machines are in use at the airport and police breathalyzers get drunk drivers off the road. Machines that can smell make the world a safer place. Several technologies are used to detect smells. One approach is to use sensor elements that change their resistance depending on the substances that are present in the surrounding air. Another approach is to measure the difference in resonance in the presence of different sized molecules, or to measure the shift in color in a reactant chemical, caused by a chemical reaction to molecular levels of substances. None of these methods offer the speed or the resolution of the nasal epithelium, the part of the nose that contains smell detector neurons.

The Sense of Taste

A human has approximately 10,000 taste buds, located in papillae on thetongue, the soft palate in the back of the mouth and the epiglottis (the leaf-like projection at the back of the mouth). Taste buds are concentrated around the tip and the edges of the tongue. Besides taste, the tongue senses touch, temperature, pressure and pain, and plays an important part in the production of speech.

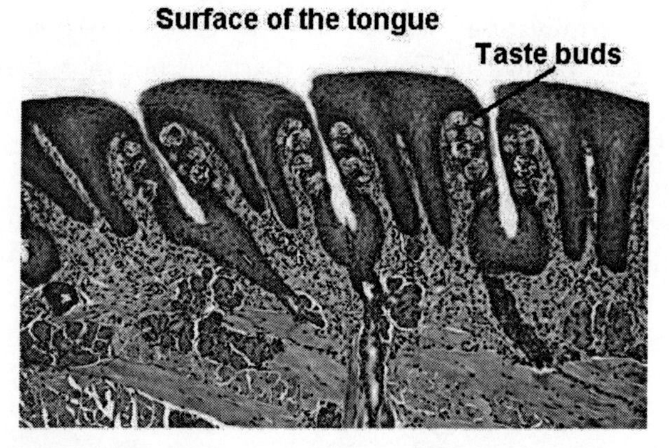

Section of the tongue showing taste buds: Image © Getty

The idea that taste is located in specific areas of the tongue is false. Sweet, salty, sour, bitter and savory taste buds are distributed throughout the mouth[25] but particular tastes may be more concentrated in specific areas. A taste bud functions similar to the chemical receptors in the nose, but its construction is different. A taste bud is an onion shaped cluster of cells that are connected to nerve endings. There are three different types of cells in a taste bud; Dark cells (up to 80%), Light cells (10-15%) and taste receptors (7-14%). There is a small gustatory pore at the end of the cluster. Food molecules are mixed with saliva and trigger specific receptors by chemical stimulation. There are four main types of papillae; filiform,

[25] Nature. 2006 Aug 24; 442(7105): 934-938. "The cells and logic for mammalian sour taste detection" Huang AL, Chen X, Hoon MA, Chandrashekar J, Guo W, Tränkner D, Ryba NJ, Zuker CS. "The cells and logic for mammalian sour taste detection"

circumvallate, foliate and fungiform. Filiform papillae are the most numerous but don't contain receptor cells. Circumvallate papillae are the largest papillae on the tongue. They detect primarily sour and bitter tastes. Each contains around 260 taste buds. Foliate papillae detect primarily sour and sweet tastes. There are 117 taste buds in each foliate papilla. Fungiform papillae sense primarily salty tastes. Each fungiform papilla contains between one to eighteen taste buds.

Artificial Taste Sensors

Can taste be measured? Taste is subjective; it is not simply the analysis of chemical compounds. A bit of salt added to pineapple makes it taste sweeter. This is known as the 'mutual effect'. Tastes differ between individuals. Smell is a large part of what we taste. When we have a nose cold we perceive much less taste. To simplify things we can concentrate on the chemical taste that is sensed by the tongue, not how the brain interprets this information. Current artificial taste sensors test for contamination in food.

The Senses of Hearing, Balance and Acceleration

These three senses are located in the cochlea in the inner ear. The external part of the ear is angled forward to catch sound from the front. The outer ear connects to the ear channel, a tube that leads to the middle ear. The ear channel is a slightly bent, 4cm long tube that contains sebaceous glands. Together with discarded skin cells and sweat, the sebaceous fluid forms the earwax. At the end of this tube the eardrum separates the outer ear from the middle ear. The eardrum is only 0.1 mm thick membrane, about a centimeter in diameter and shaped like a cone, stiff in the middle and softer towards the edges. The middle ear contains three small bones, the anvil, hammer and the stirrup. These three bones form a mechanical amplifier and the connection between the eardrum and the cochlea in the inner ear. The eardrum surface-area motion is amplified using leverage by the hammer and anvil bones, with the stirrup bone pressing into the cochlea. The middle ear has a tube to the back of the nose to allow air pressure in this chamber to stabilize. Most of us have experienced the pressure that is felt in the ear during descent when a plane comes in to land. The cochlea is shaped like a snail's shell. Unwound, it would look like a narrowing tube. In the center of the cochlea floats the cochlear partition, a membranous ribbon, which contains approximately 30,000 hair cells. These are neurons that have dendrites exposed to movement in the perilymph (fluid). Because of differing levels of stiffness, hair cells are sensitive to different frequencies. The cochlea forms a mechanical spectrum analyzer, and the brain receives signals in parallel, representing intensity and frequency from 19,000 nerve cells. The membranous ribbon divides the inner cochlea into two chambers, filled with endolymphatic and perilyphatic fluid. This forms our three dimensional sense of balance and acceleration, including the acceleration due to gravity. In free-fall this mechanism warns us of the absence of the force of gravity. Cochlear Implants are not, as many believe, an artificial cochlea. They are devices that stimulate the hair cells inside a natural cochlea with electrical pulses, thus restoring hearing to previously deaf people. People whose cochlea does not contain functioning neurons cannot benefit from a cochlear implant. Current cochlear implants are not suitable to provide hearing to an artificial brain. In an experiment, a signal generator was connected to the input of a spectrum analyzer. Output pulses that were representative of the signal generator frequency were applied to the inputs of 10 artificial neurons in a Field

Programmable Gate Array (FPGA). These 10 neurons very quickly learned to respond to ten different frequencies set up on the signal generator. It is therefore expected that Synthetic Neuro Anatomy devices will be used in future cochlear implants.

The Senses of Touch, Pain and Temperature

The skin is an elastic tissue that regulates body temperature and protects the body. It is the largest organ of the body. It provides tactile senses to the brain, which is important for tool manipulation, contact detection and surface texture. The skin consists of three major layers; the epidermis, the dermis and the subcutaneous tissue. The epidermis is 0.2 to 1.5mm thick and contains pigment cells and keratin. The dermis is 0.3 to 3mm thick and contains skin receptors, most skin nerve endings and capillaries. The subcutaneous tissue contains some nerve cells, stores energy in the shape of fat and contains large blood vessels. Millions of sensory neurons are located in the dermis.

The Sense of Proprioception

Proprioception is the sense of where our limbs are in relation to our body at each moment, what weight the bag we are carrying is and it includes our sense of balance. Proprioception is often confused, or used interchangeably, with kinesthesia. Kinesthesia is a part of proprioception. It consists of sensory feedback from the muscles, tendons and joints. Stretch receptors in the tendons, joints and muscles respond to movement by sending pulse streams back to the brain. The brain uses this information to adjust its movement of limbs and to remember where in relation to the body the limbs are. The foremost use of proprioception is to learn motion and balance.

It is responsible for 'programming' the brain modules involved in motion early in life when a child learns hand-eye coordination and how to walk.

The Sense of Time

The 24-hour cycle is controlled by a tiny organ, the size of a grain of rice, called the suprachiasmatic nucleus. It contains about 20,000 neurons and receives input from the optic nerve. This is not so much a 'biological clock' that keeps track of time as a method of regulating the 24-hour rhythm of wakefulness and sleep. No sensory organ is associated with time. How can something be perceived if there is no sensory organ associated with it? Time perception is subjective; waiting time appears to take much longer than the time that goes by when we are enjoying ourselves. The more engaged the mind is in a task, the shorter perceived time is. Time perception is a basic function found in all mammals. The removal of the cortex does not affect it. fMRI studies[26] confirm what was suspected for some time. Time perception resides in the basal ganglia, deep within the brain, but other areas also play a part. People with damage to the Cerebellum have difficulty estimating time intervals. The same is true for the right prefrontal cortex. Time perception is likely to be a result of distributed network that determines elapsed time from the sum of neural activity within the brain.

---

[26] Harrington, D.L., Boyd, L.A., Mayer, A.R., Sheltraw, D.M., & Lee, R.R. (2002). Representations of time: An event-related fMRI study. Harrington, D.L.; Boyd, L.A.; Mayer, A.R. Sheltraw, D.M.; Lee, R.R. Dept. of Veterans Affairs, New Mexico Univ., Albuquerque, NM, USA Neural Information Processing, 2002. ICONIP '02. Proceedings of the 9th International Conference on 18-22 Nov. 2002 page 423 - 427 vol.1

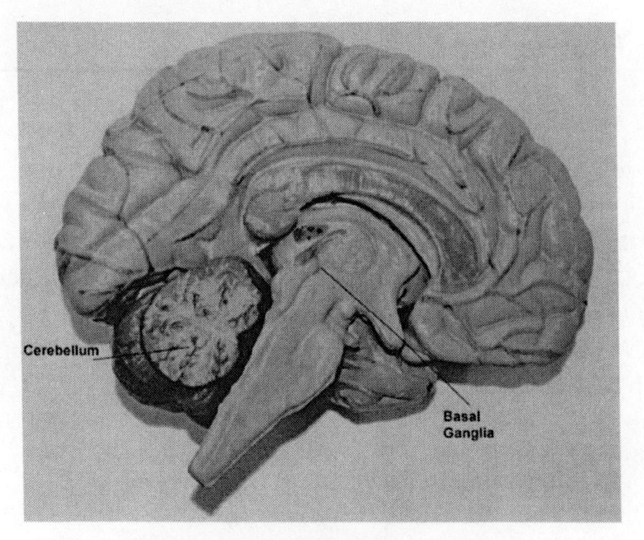

Location of the basal ganglia in the brain

Chapter 10

The Human Mind

"Some people say that computers can never show true intelligence whatever that may be. But it seems to me that if very complicated chemical molecules can operate in humans to make them intelligent then equally complicated electronic circuits can also make computers act in an intelligent way. And if they are intelligent they can presumably design computers that have even greater complexity and intelligence."

Stephen Hawking, at the White House on March 6, 1998.

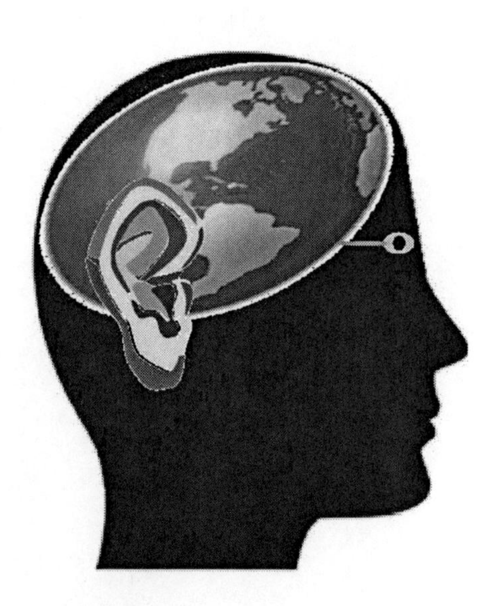

The mind inhabits the brain, and it gives birth to our ideas, our philosophies, our beliefs, our personality and our fears. It is the seat of our being and our intelligence. The mind is who we are and what we know, or think we know. It is the 'program' that forms and enables the brain to accomplish its function. The mind is dependent on the anatomy of the brain, just like the brain anatomy is determined by the mind. Without the brain's structure, density and plasticity the mind could not develop, e.g. a 'small' network of a million or so neurons and say 10 billion synapses is not likely to develop a mind with the complexity of a human. There are not enough storage locations and processing nodes to accommodate the human mind. A cockroach brain has about a million neurons and it has very little personality. Whole libraries of books have been devoted to psychology, the science of the mind, but here we need to limit our scope to what is relevant to a synthetic brain. Will a synthetic brain of sufficient complexity, constructed from columns of artificial neural and glial cells develop a personality, and what can we expect that personality to be? Much will depend on how well its structure emulates the anatomy of a real brain and the training that the synthetic brain receives. This could give birth to a whole new discipline of synthetic brain psychology.

Mind Uploading

Mind uploading becomes achievable when it is possible to read the chemical composition of a hundred trillion synapses and the complete connectome of a living human being. At the same time, a medium such as a synthetic brain is necessary to contain the mind. It can be assumed that a synthetic brain that contains an uploaded mind, retrieved from a living person's brain, will reflect the personality, consciousness and the beliefs of that person. The brain would need to be 'tuned' to contain this mind. It would require the correct neuron (grey matter) densities throughout specific areas of the brain, as well as a means to connect any processing node to any other node to reflect the connectome of the person. Brain structure is a factor of the knowledge that is stored. Significant parts of the nervous system would need to be copied, including the limbic system, the brain stem, the spinal cord and the sensory organs. The inner brain and sensory system support the function of the cortex; a cortex cannot exist in itself. It would not perform any function since it would be unable to receive sensory input and therefore cannot learn anything. Mind uploading is a philosophical topic that must remain a notion for the future for now, since we don't yet have the technology to scan a human brain at the molecular level, which is necessary to 'read' the density and chemical composition of neurotransmitter chemicals. For now it appears to be feasible to allow a new mind to develop in a synthetic brain. To explore the idea of a synthetic mind that resides in a synthetic brain further we will have to refer to the mind that is stored in our own brain. The human mind consists of the conscious, subconscious and unconscious mind. The conscious mind is where we do our thinking, where we are aware of our actions, our behavior, judgments, and observations.

The unconscious mind also contains our belief systems. These belief systems have been established over the years, first by our parents, then by our teachers and peers. We are not directly aware of them, because they reside in the unconscious mind, but they do rule our life. Belief systems have immense power over our concept of who we are and what we are capable of. Since we all have similar brains, we each should be capable of anything another person can do. It may take one individual a little longer than another. Henry Ford, the US industrialist, said: "Whether you think you can or think you can't - you are right."

Our mind has immense power over our body, which includes the brain itself that contains the mind. The mind controls hormone levels. Contrasting beliefs in the conscious, subconscious and unconscious mind causes heightened stress levels. Everyday living causes a further increase in these stress levels. This becomes a problem when the mind is keeping the body in a 'fight or flight' mode over a prolonged period of time. The muscles are tightened, blood pressure increases and adrenaline is released into the bloodstream, increasing blood sugar and oxygen levels. All this was originally intended to make us either run faster, or to fight more ferociously.

Artificial Brain and Mind Development

When we train a synthetic brain, as we would train a child, then the artificial mind will reflect all aspects of that training. There is a great debate between supporters of the nature versus nurture question. This will present us with an opportunity to examine this question and experiment with it like we have never before been able to. No one knows how much of our mind is the result of training and how much determined by our genes. This will be one of the many exciting research avenues that are opened up by the event of an artificial brain. A machine with the complexity of the human brain will be impossible to train without some low level structures in place. These low level structures will be developed in small synthetic brains, that although small, have the same structure and layout as the much larger synthetic neuro-anatomy systems that emulate the human brain. On these smaller systems a library of unloadable "Training Models" is developed. These training models are stored in a library and combined to give the synthetic brain the same advantages that a human baby has; the ability to learn from its environment. The synthetic brain will need senses to interact with its environment. It will need muscles and feedback from those muscles to know where its limbs are. From this feedback it develops the ability to stand up, and to walk. From artificial eyes it learns to recognize shapes, and to associate those shapes with words. It learns words through an artificial cochlea, and produces sound by passing 'white noise' (random sound waveforms that sound like a hiss) through filters that simulate the throat, nasal cavity, tongue, lips and vocal cords. The sound it produces is received by the artificial cochlear and forms a feedback path that aids in the training of the synthetic brain to produce and understand human speech.

The Human Mind, According to Freud[27]

Conscious Mind

Higher Consciousness	
	Aware of own motivation.
	Behavior determined by choices.
	Knows the difference between own motivation and conditioning.
	Conscious awareness of truth.

[27] The Standard Edition of the Complete Psychological Works of Sigmund Freud

Normal Consciousness	
	Observes, analyzes and judges information based on conditioning.
	Motivated by feelings, thoughts and desires.
	Unreflective.
	Consensus.
	Reactive behavior determined by social forces.
	Behavior determined by the environment.
Pre-consciousness	
	Assimilation of feelings, ideas, and thoughts before awareness.
	Old experiences applied by relevance or similarity (not necessarily appropriately applied).
	Repressed hurts, feelings and thoughts.
	Acquired defense mechanisms.
Sub-conscious Mind	
	No awareness to the individual.
	Reactive reservoir of circumstances of the past.
	Insertion of new information and knowledge.
	Determines patterns of behavior.
	Emotional and cognitive experiences of the past.
	Phobia.
	Obsessions and delusions.
	Compulsion.
	Suppressed memories of painful events.

Unconscious Mind	
	No awareness to the individual.
	Fundamental survival drives.
	Bodily sensations.
	Fundamental beliefs.
	Deep buried emotions and repressed feelings.
Personality	
Identity	No direct awareness to the individual.
	Driven by pleasure principle, instinctive drives.
	Passions.
	Seeks immediate gratification without morals.
	Fundamental wants and needs.
	Seeks to avoid painful situations.
	Seeks to be creative.
Ego	Defers instant gratification because of reality.
	Self-image.
	Seeks to satisfy the needs of the identity
	Reason and common sense.
	Driven by the identity, social acceptance, and superego.
Super-Ego	Rules for good behavior.
	Social acceptance.
	Internalized moral standard.
	Behavior models acquired from parents.
	Pride, value, feelings of accomplishment.
	Feelings of guilt, inferiority.

Constructing a mind that operates a synthetic brain is going to take many small steps. It is a huge step from a first synthetic neuro-anatomy device with barely 15,000 nodes and 240,000 synapses to the emulation of the 86 billion nodes and 100 trillion synapses of the entire brain. Many intermediate steps are necessary. When all of these are built upon the same platform technology then the results can be carried forward.

If we succeed in scanning the mind of a living person, we will also need a synthetic brain that has the correct neuron densities and synapses to contain that mind. These two approaches, the development and training of artificial brains, and the creation of artificial brains as a medium to contain a human mind, can progress side by side and benefit from each other. As an added bonus we will have some ideas about the interpretation of information that is received from a living person's brain.

Chapter 11

Animal Intelligence

"Surely man is the king of the beasts, for his brutality exceeds theirs…"
Leonardo da Vinci

The sphinx is a lion with a human face.

The human brain with all its complexities may be too large to emulate. Animal brains are simpler, but will a smaller brain automatically mean 'dumber'? A smaller brain will not have the storage capacity of a larger brain. Much depends on structure. Imagine for a moment not having language in any form, not being able to plan our day, or to read and 'hear' the words in our mind. Children who grow up without language, due to physiological defects or social circumstances have difficulty grasping simple ideas, to the point where it was once thought that they were mentally handicapped. Without words, animals think differently from humans. They experience feelings, impressions, and memories of events and act on experience and instincts. They have a limited set of noises that indicate danger, warn off competitors and enforce social status etcetera, which are somewhat similar to our non-verbal communications of grunts and nods. Anyone who lives with a teenager knows what I mean. People who work with animals are generally able to interpret if the animal is in pain, aggressive, afraid or happy. What can we learn about intelligence from the brains of animals, and how does this relate to machine intelligence? Much research has been done on animal brains. Animals are sentient. They are aware, feel loved or rejected, feel pain, grief, fear and pleasure. Many animals are used as unwilling participants in research as test subjects. Some of these research projects involve surgically inflicted damage to the subject's brain to study its effects. Test subjects are killed and their brains are dissected.

An average human's brain weighs almost 1400 grams, with men's brains on average 10% larger than women's brains. Active brain mass that contains neurons, or 'gray matter', is distinct from passive brain mass or 'white matter'. Grey matter uses a lot more energy than muscle tissue by a factor of sixteen times[28]. Hence, more intelligent species should consume more energy by body size than simpler species.

Species	Neurons	Average Brain Weight (grams)
Human	86 Billion	1400
Dolphin		1550
Orca		5600
Elephant	200 Billion	5500
Cat	1 Billion	30
Dog	1.1 Billion	72
Parrot		
Sperm whale		7800
Chimpanzee		420
Octopus	500 million	

Table compiled by the Author from a number of sources

The brain to body ratio of an average human is 1 to 40. An average mouse has the same brain to body ratio. A large but fit man may have a brain weighing as much as 1600 grams and a body weight of 120 kilogram's, resulting in a brain/body ratio of 1 to 75. Brain size is obviously no measurement of intelligence. A somewhat more refined measurement than body to brain size ratio is called the 'Encephalization Quotient', in which the brain to body ratio is corrected with a factor that is specific for each group of species. For mammals this constant is 0.56 according to Kuhlenbeck[29]. Still, this does not give an accurate measurement of intelligence. Body to brain size is only

---

[28] Aiello LC, Wheeler P. The expensive-tissue hypothesis: The brain and the digestive system in human and primate evolution. Curr Anthropol. 1995;36:199–221

[29] 1973 Kuhlenbeck, H. "Central Nervous System of Vertebrates". Vol. 3, Part II, New York, NY: Arnold-Backlin-Strasse

one factor as is the quantity of grey matter (nerve cells) to white matter (myelin). Brain structure, how well sections of the brain communicate, and the density of grey matter are all factors that determine intelligence ability. Training and determination are factors in the development of potential intellect.

Tiny Brain but Smart

African Grey parrots have tiny brains compared to humans, but have been shown to associate words with meaning. Parrots brains have particularly well developed areas for speech and communications. They can solve linguistic processing tasks as well as a child. Dr. Irene Pepperberg[30] trained a parrot named Alex to recognize shapes, colors, and to count the number of children's play blocks. Alex accomplished these tasks and communicated with his trainer with an amazing level of understanding, Alex reached a high level of sophistication over a period of 22 years of intense interaction and dedication. Alex did not just repeat the words of humans, he responded to questions in a manner that indicated intelligence, reasoning and choice. Unfortunately he died on September the 6th 2007 aged 23. Human intelligence largely resides in the upper layers of the brain, called the cortex. Parrot brains do not have such a structure. Their brains are organized differently from humans. While the cerebellum and the pons are recognizable, a structure not found in the human brain is present, called the pallium. It consists of three parts; the hyperpallium, the mesopallium, and the nidopallium. Its function appears similar to the human prefrontal cortex.

---

[30] Harper © 2008 Alex & Me: How a Scientist and a Parrot Uncovered a Hidden World of Animal Intelligence — and Formed a Deep Bond in the Process, by Dr Irene Pepperberg ISBN 978-0061672477

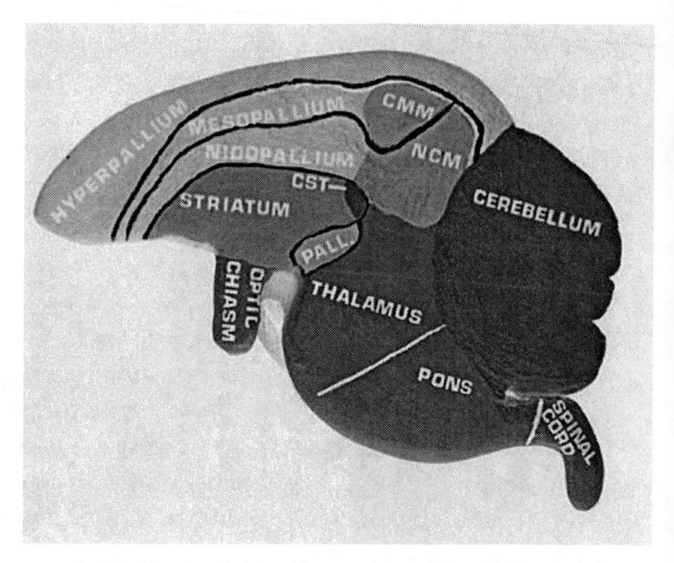

The structure of a bird brain. Original image from: Nova Science Now.

These are the areas where a parrot performs its language processing and reasoning. Given that parrot brains are the size of a large walnut, their intelligent behavior is likely to be a matter of neuron density and brain structure.

Primate Intelligence

Apes appear to be closest to humans in intelligent behavior, but their brains are about a quarter to a third the size. Lack of language ability makes it difficult to communicate with these animals to determine how intelligent that they really are. They are physically unable to produce human speech, as they do not have the equivalent of human vocal cords. Attempts have been made to teach apes sign language. While they can learn to apply the signs for about a 1,000 words, their hand gestures are sloppy, but they clearly understand and respond to the trainer's hand gestures. Chimpanzees have the ability to observe and learn. These apes have been observed in the wild to make tools to go hunting and to suffer grief when a baby dies. They lack control of their emotions, lack any loyalty to their peers and lack the ability to pass complex knowledge on, either in a written form or verbally. A chimpanzee exceeds human ability in short-term memory, as shown in research by Dr. Sally Boysen[31]. In a numbers game, the numbers one to nine are flashed on a screen for one second in a random location. A chimpanzee is able to press blank squares, where the numbers have been, in the correct order 100% of the time. The best that humans can do in an equivalent test is 30-40%. The organization of an ape brain is similar to the human brain, but with very discernible differences in the frontal cortex and in verbal processing. Their intellect is different from human intellect, too different to make a simple comparison expressed as a number.

Relevance to Artificial Intelligence

All this gives us a clue that an Artificial Intelligence system that is a smaller version of the human brain will not behave as a slightly 'dumber' entity, but as a different entity. Artificial Intelligence systems consisting of artificial neurons will, through necessity, have to start small. A chip with the complexity of a Pentium processor could contain up to 15,000 neurons and 300,000 synapses. A complete 8 inch wafer, using wafer scale integration, could contain 150 million neurons and 3 billion synapses. This is almost equivalent to the number of neurons in a dog's NeoCortex.

---

[31] Dr. Sally Boysen. Comparative Cognition Project at Ohio State University

Chapter 12

A.I. in Robots and other machines

"Technology is nothing. What's important is that you have a faith in people, that they're basically good and smart, and if you give them tools, they'll do wonderful things with them."

Steve Jobs

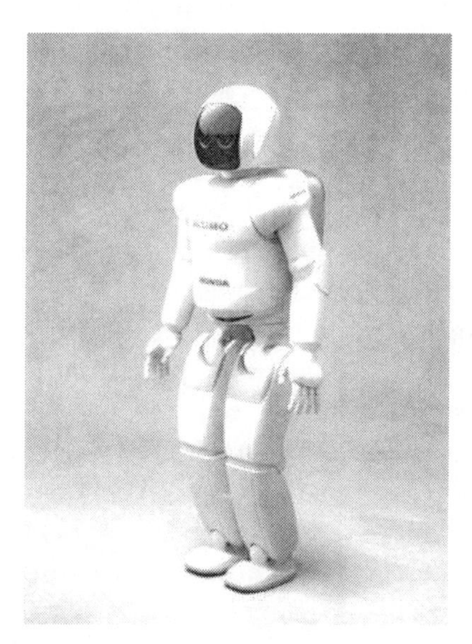

ASIMO robot © Honda Motor Co. 2011

New technologies that are heralded as being 'the next new thing' that is going to revolutionize the A.I. industry, appear every so often. There was the 'Wafer Scale Integration' neural network by Hitachi, and the CAM-brain by Hugo DeGaris, among many others. Far reaching statements were made. But did they have a clear understanding of what an intelligent system really should be, or what technology would be required to create intelligence? Systems that receive input from sensors and then use a program to take a predefined action are not intelligent, they are control systems. Likewise, systems that exist out of elaborate matrixes that simulate the selection process of neurons, but lack the ability to learn, are not intelligent. Robots, no matter how fast they are and how lifelike they appear, are largely programmed dolls. A human programmer defines all movements, capabilities and aspects of their behavior. Even within these limitations, significant computer resources are required to make a robot perform its functions. Newspaper stories about the simulation of an entire brain must be considered with some skepticism. What comprises an entire brain, and how well are its cellular functions and its structure emulated? The key is whether the emulated brain is capable of learning from sensory input. It is not enough to show that the frequency and activation patterns are similar to a biological brain. Large supercomputers are required to simulate intelligent behavior that includes learning and applying learned knowledge. The IBM's Blue Gene/Q, China's NUDT Tianhe-1A, and the Fujitsu K supercomputers are capable of simulating between 5 to 12% of a human brain. Such computers consist of thousands of processors. Each processor has memory and I/O, and thus is a 'Von Neumann' computer. A few years ago there was a controversy about IBM's claim to have simulated all of the neurons in a cat's brain, 1 billion neurons and 10 trillion synapses, although at $1/100^{th}$ of its speed. Supercomputers perform up to 10,000,000,000,000,000

(10 petaFlops) floating-point operations per second. Impressive numbers, but this is just math and not intelligence as we know it. An abacus can perform calculations, although not very fast. Like an abacus, these machines need a human hand to write a program before they perform any function. The number of calculations that are required to simulate the human brain's functions are prohibitive. There is much doubt whether human intelligence can be expressed as a series of very fast calculations. In none of the reports of such simulations is there any mention of sensory perception or learning.

ASIMO. Honda Motor Co. Japan

Over the years, Honda has received media coverage with their ASIMOrobot. This machine has been around since the mid-eighties in various incarnations, with the earliest models resembling a box on legs, which was attached to a large computer and controlled by a human operator.

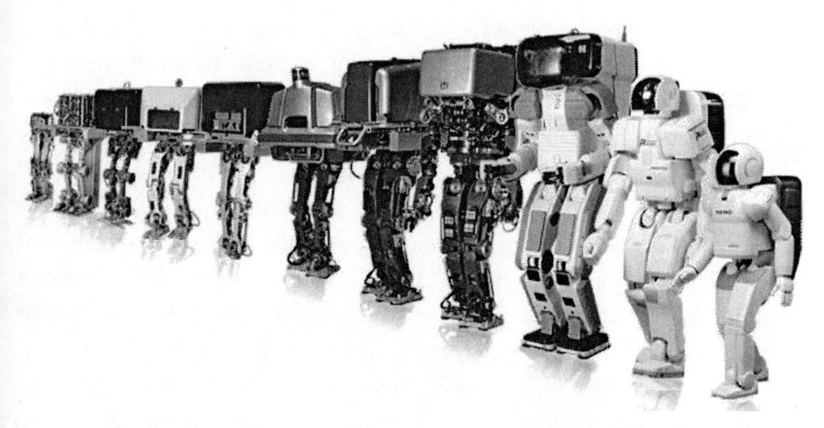

Honda © ASIMOprogrammed robot generations.

The focus of the project has been to develop stable 2-legged mobility, a natural walking ability that allows the robot to climb steps and move over uneven ground. The latest version is clad in white plastic with a black faceplate. Any physical resemblance to a storm trooper may be coincidental. This is an impressive effort of mechanical and electronics engineering. It stands 1.30 meters tall, and weighs just 54 kilograms (4 ft 3in.and 119 lbs) due to its aluminum alloy and plastic construction. Its backpack stores a fifty-volt Lithium Ion battery that can power the robot for about an hour. The on-board robot controller communicates with a remote control computer over a wireless Ethernet link. The walking algorithm is the result of more than 25 years of research and development into the physics and the anatomical processes of 2-legged motion. ASIMO can walk, run, balance and play with a football. It takes into account people's movement and predicts where they are heading to avoid a collision. It is more dexterous and able to grasp soft objects such as paper cups, without crushing them. Behind the face mask a camera enables the robot to perform face recognition. It also has the ability to perform speech recognition and speaker identification using an array of eight microphones that are distributed around its body and in its head. This method is intended to cancel out background noise and to make it possible to analyze more than one conversation at the same time with multiple people talking. In real-life situations, it recognizes about 30% of what is said in noisy, multi-speaker environments.

Dynamic Walkers

As an alternative to the complex, controlled way of walking used in ASIMO, Passive Dynamic walking is an efficient motion technique pioneered by Tad McGeer and Eric

Vaughan[32]. It's based on the mechanics of human motion. Unlike ASIMO, it involves no computer or electric motors. A passive dynamic walker can walk down a slope using only gravity. A powered adaptation of this machine can walk over level terrain and uses very little energy, much less than what is used in a walking machine with multiple electric motors. Studies have shown[33] that neural learning algorithms can be used in combination with the dynamic walker method to evolve walking robots. Dynamic walkers, controlled by a dynamic temporal associative memory that learns to control the legs by feedback, represent a novel path forward in walking robot development.

WATSON. IBM Research. USA

Watson is an IBM supercomputer consisting of 10 refrigerator-sized racks, dedicated to playing 'Jeopardy'. It performs this feat by Natural Language Processing (NLP), which provides the key words for a fast relational database search engine. The questions are displayed on a monitor and read out by the host of the show. The monitor text is also input to the Watson computer. This gives the computer a slight advantage. It does not 'read' the text, but straight away starts to analyze it and process an answer. The computer exists out of a cluster of ninety Power750 servers with some special network, clustering and I/O devices. At 80 Teraflops (80,000,000,000,000 floating

---

[32] The Evolution of Control and Adaptation in a 3D Powered Passive Dynamic Walker (2004) Proceedings of the Ninth International Conference on the Simulation and Synthesis of Living Systems, ALIFE'9. Passive dynamic walking (1990) by T McGeer, The International Journal of Robotics Research

[33] Artificial Life 2004 © MIT. The Evolution of Control and Adaption in a 3D Powered Passive Dynamic Walker. Eric D. Vaughan et al.

point operations per second) it comes in at a slow 94th in the list of top 500 supercomputers. Watson has a total of 2880 processor cores and 16 Terabytes of memory, running IBM's parallel processing DeepQA™ software. A single PowerPC processor would need at least 2½ hours, but Watson's multi-processor cluster is designed to return an answer within 3 seconds. In this time it searches the entire database, containing millions of documents, including Wikipedia texts, dictionaries, newswire articles, encyclopedias and literary works, etc. To increase performance, the database is kept in local memory, and not on disk. Natural language processing was first considered in a 1949 memo by Warren Weaver[34]. In the context of machine translation he considered four points:

a. Multiple meanings, which can be solved by looking at the word within the context of the sentence.
b. Logical syntax elements and semantics of language including the formal rules and theory of the language.
c. Application of cryptography methods to machine translation, e.g. a Turkish document is just an English document coded in Turkish.
d. Like point (b), but applied to linguistic universals; words that are commonly used in combination to express meaning.

Watson deciphers a sentence by using an advanced set of rules that are based on Warren Weaver's work. This provides a set of search criteria, which are then filtered to avoid searches of common 'filler' words that are not relevant to meaning. The question is then parsed for a word that describes the answer and the properties of the answer, e.g. is it a person, an

---

[34] Warren Weaver Memorandum, July 1949. Natural Sciences Division, Rockefeller Foundation

organization, a company, a place, a book, a movie or a physical object. In the next step some semantic assumptions are generated, such as, is the person an author, an actor, a character in a book, or a celebrity? Database searches are built based on the first two processing steps. The database, equivalent to some two million books, is searched next. This generates hundreds of possible answers. Evidence is generated and new assumptions are made. The answers are then scored for the greatest consensus with the assumptions that were made in the first two steps, and weights are assigned to each possible answer. A 'confidence' value is computed, which links back to the evidence. The buzzer is pushed once an answer is determined, which is then synthesized as human speech. It beat both human Jeopardy champions, but what does this prove? Watson's program is a leap forward in the searching of database data using natural language. But just like a chess-playing computer program, it beats its opponents with logic, and the ability to perform fast database queries. The answer is computed data without awareness. Not intelligence.

Brains in Silicon

The Brains in Silicon project is a group at Stanford University. The group's objectives are to build an affordable supercomputer and to gain a fundamental, biological understanding of how the human brain works. Both objectives come together in the research of neural functions and connectivity. The brain is a supercomputer like no other machine. Designing at the transistor level, the Brains in Silicon group has created an analog neural processing chip that uses electron migration in MOS transistors to emulate synapses[35].

---

[35] Kwabena A. Boahen, A. G. Andreou, et al Architectures for Associative Memories using Current-Mode MOS Circuits, Advanced Research in VLSI, C L Seitz, Ed, MIT Press, Cambridge MA, 1989.

Each neuron is mapped to an address, which can be dynamically connected through a digital multiplexer. The address points at a memory location that holds the synapse's strength value that is supplied back to the chip for integration. A number of concept projects have been completed; an artificial retina[36], an artificial cochlea[37] and functional models of sections of the brain including the hippocampus[38], and the basal ganglia.

IBM SyNAPSE Project

The IBM "Systems of Neural Adaptive Plastic Scalable Electronics" project is aimed at building devices to emulate the brain's computing efficiency and compact size, without being programmed. This is accomplished by emulating the brain's Synaptic Time Dependent Plasticity (STDP) to facilitate learning in much the same way as the method proposed here in previous chapters, except that the IBM team deemed it necessary to develop the synapse 'training patterns' offline in a software simulation[39] to upload to the chip. The chip uses memory to store synapse parameters such as neurotransmitter type and strength. It is intended to speed up neural processing

---

[36] K A Zaghloul and Kwabena Boahen, A silicon retina that reproduces signals in the optic nerve, Journal of Neural Engineering, vol 3, no 4, pp 257-267, December 2006

[37] Wen B. Modeling The Nonlinear Active Cochlea: Mathematics and Analog VLSI. Doctoral Dissertation, Department of Bioengineering, University of Pennsylvania, Philadelphia, PA, 2006.

[38] Arthur J. Learning In Silicon: A Neuromorphic Model of the Hippocampus. Doctoral Dissertation, Department of Bioengineering, University of Pennsylvania, Philadelphia, PA, 2006

[39] "A Digital Neurosynaptic Core Using Embedded Crossbar Memory with 45pJ per Spike in 45nm". Paul Merolla, John Arthur, Filipp Akopyan, Nabil Imam, Rajit Manohar, Dharmendra S. Modha. IBM Research - Almaden, 2Cornell University

and it incorporates many of the features of the technology presented here.

EPF Blue Brain project

The blue brain project is run by Prof. Henry Markram at the École Polytechnique Fédérale de Lausanne (Federal Polytechnic School at Lausanne, Switzerland). The project aims to simulate the human brain in software on an IBM Blue Gene supercomputer. The IBM Blue Gene is a scalable supercomputer consisting of up to 64 fridge-sized enclosures and consumes enough power to supply a small town. Henry Markram said in 2006, in Nature[40],"Computational power needs to increase about 1-million-fold before we will be able to simulate the human brain, with 86 billion neurons, at the same level of detail as the Blue Column." The model is based on Michael Hines's NEURON[41] software, which allows all known aspects of a biological neuron, such as temperature, axon myelination, channel types, membrane densities, and the ion concentrations to be programmed. The program calculates the neuron output voltages over time. The Blue Brain Project consists of a large matrix of such neurons, their synapses and interconnection organization. The aim of the project is not Artificial Intelligence, but to build an accurate model of the human brain.

Singularity Institute and AGI

Ben Goertzel, one of the advisors of the Singularity Institute, has been promoting the term "Artificial General Intelligence",

---

[40] Perspectives, Nature. Page 160, February 2006, Volume 7

[41] Michael Hines. Yale University. Supported by NINDS grant NS11613 and in part by the Blue Brain Project.

or AGI, to set it apart from Artificial Intelligence (A.I.). He says that it was coined by Shane Legg, an A.I. researcher who once worked for him. The name of the institute is based on a book by Ray Kurzweil "The Singularity is Near"(2005) in which he offers many predictions, including the opinion that the singularity – the event when computers become more intelligent than man – will occur within the next 25 years. This is to be accomplished by computers that are millions of times faster than our present-day computers. He expects a computer that can contain the entire human brain to be available by 2030. This means a PC that is a hundred million times faster than 2010 technology, and with at least 3 Terabytes of memory. The research areas of the Singularity Institute include the formalization of the Artificial General Intelligence project, the ethics of Artificial Intelligence, "Friendliness Theory" and systems design. The Singularity Institute organizes the yearly Singularity Summit that brings together hundreds of A.I. researchers and interested parties.

Groundbreaking Materials Research

Carbon nanotubes are cylindrical, molecular sized, graphite carbon tubes with a lattice-like appearance. They are strong, lightweight and excellent conductors. Their electrical characteristics are better than copper. In the future they could replace the metal layers in integrated circuits, or whole microcircuits could be built from carbon nanotubes in combination with some type of dielectric material. Researchers at several US universities have succeeded in creating a field-effect transistor using carbon nanotubes and zirconium oxide that could revolutionize the semiconductor industry. However, the path from a small-scale research component to full production environments is long and difficult, with many stumbling blocks on the way.

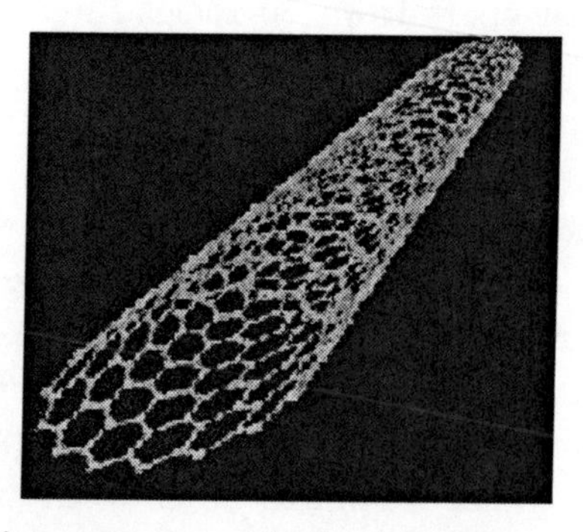

Lattice structure of a Carbon Nanotube

Take bubble memory for instance. In the late 1970's hundreds of millions of dollars were invested in the production of bubble memories. The science behind them was solid, production prototypes worked according to specification, but the factories could never get more than one in ten to work. In some cases the yields were so low that only one device in a hundred worked. Then flash memory appeared and that was the last nail in the coffin of bubble memory. Huge factories were decommissioned overnight, and giant losses were written off. The manufacturing difficulties surrounding carbon nanotube processors are reminiscent of the early days of bubble memory production. "These problems will be solved," they said. When they weren't, the technology eventually died. NanoWerk is a nanotechnology company located in Hawaii. Dr. Krupke said, "It should be possible to scale up the fabrication technique to simultaneously and reproducibly fabricate a very large number of such devices on a single chip, each accessible individually for electronic transport. Conventional nanotube growth and device fabrication techniques using chemical vapor deposition or spin-casting are unable to achieve this, due to a lack of precise control over nanotube positioning and orientation." The environmental and health risks imposed by carbon nanotubes are also a concern. There is mounting evidence[42] that carbon nanotube pollution in waterways will cause algae to die off. The effects on the human body may be similar to the dangers of asbestos fibers[43].

Quantum computing

Quantum computers are fundamentally different from 'standard' computers. There is some debate whether quantum computers actually exist beyond pure theory. Binary 'bits' have two states, on or off, also referred to as True or False and One and Zero. In traditional logic an ON state means that there is +5 volts present, while an OFF state represents zero volts. A third state exists that is not really a 'state', but the absence of any level. All the wonderful things that computers do these days are performed using just these two states. All data and the processor instruction set are encoded as sequences of bits. In Quantum Computing, Qubits are used. Each Qubit has multiple simultaneous states. Quantum computers are nothing like regular computers. They work in ways that are completely different. The strength of Quantum computers lies in the fact that they can solve problems that are impossible to solve, or simulate, on a conventional computer. A quantum computer consists of single atoms, and each atom represents a single qubit, suspended in a magnetic field. Four qubits are together about the width of a human red blood cell. In a 4-qubit

---

[42] Fabienne Schwab, Thomas D. Bucheli, Lungile P. Lukhele, Arnaud Magrez, Bernd Nowack, Laura Sigg, Katja Knauer. Are Carbon Nanotube Effects on Green Algae Caused by Shading and Agglomeration? Environmental Science & Technology, 2011

[43] Craig A. Poland et al. Nature Nanotechnology 3, 423 - 428 (2008) "Carbon nanotubes introduced into the abdominal cavity of mice show asbestos-like pathogenicity in a pilot study".

computer, there are 2^4 numbers. In a 10 qubit computer there are 2^{10} numbers and $2^{2^{\wedge}10}$ combinations. Laser light is used to perform logic operations and to 'read' the qubits. These operations affect just one qubit, but cause a whole sequence of changes that ripple through the other qubits. To understand this process, it is necessary to delve into quantum mechanics. There is an excellent book by Richard Feynman on this subject called "QED: The strange theory of light and matter" Princeton University Press. ISBN 0-691-12575-9.

Subnano and Multicore Silicon

People have been speculating over the last 13 years what sort of microprocessors would be available by now. Forecasts made in 2000 ranged from a 10 GHz processor core made by Intel, to wild speculations of a 380 GHz Nanotube processor by optimistically accelerating Moore's Law. So far the forecasts of a 10 GHz microprocessor have not materialized, although quad-core processors nearly reach the same performance. For now, they seem to put their money on multi-core silicon, with processors forecast that predict 100 or more cores on a single chip. Silicon processor chips will be around for some time to come. All the silicon foundries in the world are tooled up to make silicon chips, which represents a huge investment. The individual core processors will be simpler in design, placing a higher priority on parallel processing than on processing optimization.

Chapter 13

Computer Control, Programming and Learning

"To approach the stranger is to invite the unexpected, release a new force, let the genie out of the bottle. It is to start a new train of events that is beyond your control..."

T.S. Eliot

Traditional microprocessor-based control systems use a wide variety of sensors to interpret the outside world. Such sensors generally generate a voltage that represents a single entity, such as temperature, pressure, proximity, fluid level, frequency and other physical phenomena. The generated voltage is typically very small, in the range of a few thousandths of a volt. This miniscule voltage is amplified and then converted into a binary value. This is called an ADC, an Analog to Digital Conversion. The binary value tracks the sensor voltage at regular intervals. The binary value is input on a port, and is visible to the program as "labeled data"; the binary value is directly representative of the level of the physical event. The programmer knows the meaning of the binary value that enters the system at a particular port, and also is aware of the level it represents. An example of a large control system is the automated car park system installed at an exhibition center that received input from over 500 sensors, and controlled nearly 500 lights, gates and ticket machines. This was done long before the PC existed. It took weeks to label all the wires, identify the ports that they connected to, and to debug the values received by the control programs.

Computer Architecture and History

A computer's only function is to fetch program steps from memory and to perform logic operations; it has just as little intelligence as a tape recorder. No one would consider a MP3 player to be a great composer when it plays back a symphony. A voice recorder has obviously no awareness of the meaning or context of a recorded speech. Likewise, a computer plays back the programmer's recorded intellect in a processor 'mill' that fetches program instructions one by one. When a computer wins a chess game it is the human programmer who won the game, aided by the arithmetic and logic of the computer.

All of today's PCs are 'von Neumann' computers. Von Neumann described the stored program computer structure in the 1940's. A central processing unit (CPU) retrieves program instructions and data from memory. The CPU is a 'mill' that constantly goes around in a loop. This loop consists of fetching a program step, fetching the relevant data from memory, performing the function that is expressed by the program step, and storing the data in memory. Then the next program step is performed in the same way. Each program function is performed by decoding the program step into microcode. The microcode directly controls the Arithmetic Logic Unit (ALU) within the microprocessor. To illustrate this process consider an ADD X, Y instruction. To execute this instruction the microcode is:

Load register X into ALU latch A
Load register Y into ALU latch B
Set ALU in XOR mode
XOR latch A with Latch B
Store the result in register X
Set FLAG register to indicate carry (overflow)
Set FLAG register to indicate zero result
Increment the Program Counter

The ALU performs logic operations on data in processor registers and memory locations. It can also perform an addition operation on the program counter register, causing the program to jump to a new location. All behavior is governed by its programs, which are long lists of instructions, like ADD, SUBTRACT, JUMP, COMPARE and MOVE. Everything is coded in ones and zeros. Simple 2-state yes-no decisions are made by subtracting the values in two registers and examining the sign of the result in the Flag register. Programs on a PC exist within a hierarchical structure. Program routines are layered on top of simpler routines, like the layers of an onion (please refer to the image on the next page). At the bottom of this hierarchy exists the machine components themselves, with the BIOS (Basic Input Output System) program interacting directly with hardware. The BIOS exists out of machine code that handles simple tasks such as sending a single character code to an output port, and receiving keyboard characters into a memory called the keyboard buffer. Right above this layer exists the HAL or Hardware Abstraction Layer. The HAL contains routines that make it possible to communicate with different manufacturer's components that perform a similar function by translating their control codes to a standardized interface. The HAL supports the Windows Application Program Interface or API. The Windows API contains functions to handle programming threads, to create a graphics window on the screen, to create a tool bar, to draw or display graphics icons in a specified location, and to handle disk files. An application program selectively calls functions in the API. Each API routine calls simpler functions in the HAL, which calls functions in the BIOS, which controls the machine hardware. Although it has seen many updates over the years, the BIOS is the oldest part of this hierarchy, preceding the existence of Microsoft. This hierarchy of API and library functions is by no means unique to Microsoft Windows™. It has existed for a long time in software architecture, ever since the early days (1969) of the UNIX operating system. Reusability of simpler functions has shaped the evolution of software into more and more complex programs. The Windows user interface, the thing we see on the screen, is one of those application programs. There is a device driver that 'reads' the input from the mouse. This value is available to the user interface program through an API, and the program then calls another API to draw the arrow in the appropriate position on the screen. The whole Windows operating system consists of many modules, some of which have changed little since the days of Windows NT ™.

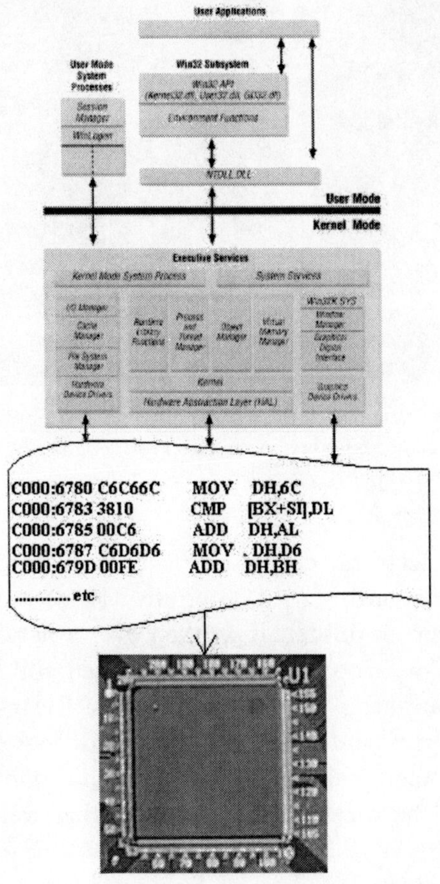

Microsoft Windows ™ processing method

A Computer Family Tree

ENIAC was one of the earliest electronic computers. This image was taken in the late 1940's. (US army image, now in the public domain)

The earliest electronic computers were built from thousands of relays. The Harvard Mark I computer (1944) was the first US-built electronic computer. It worked over a period of 15 years. In Germany two young engineers designed and built the Zuse Z1 to Z4 computers from 1936 onwards. Relays were used to create logic gates, and the effect was an insistent clicking "like a thousand women knitting". The electronic computer was an evolution of the mechanical calculators that were in common use then and many of its mechanisms used the same methods, which go back to the days of Babbage. An electronic circuit replaced the gears of the mechanical tabulating machines. Program instructions were short sequences of ones and zeros, in which each 'one' could be traced to a specific action of the machine. The earliest computers were programmed by manually throwing switches and loading the resulting value into memory. Individual switches were replaced by punch tape. The punch tape was a long roll of 1-inch wide, tough paper with holes punched in it. This tape was fed through an electro-mechanical reader consisting of micro-switches. A hole caused a closed contact resulting in a '1' while a open contact was a '0'. The relays of the first electronic computers were replaced by valves and transistors in the newer models. By the late 1960's the first DTL (Diode-Transistor Logic) "5400" series of integrated circuits were used, which were soon superseded by 7400 series of TTL (Transistor-Transistor Logic) circuits. The SN7400 contained four NAND gates. Mainframe computers contained thousands of these integrated circuits on small plug-in boards. The boards plugged into a wired backbone, thousands of color-coded wires and connectors in a 19" equipment rack.

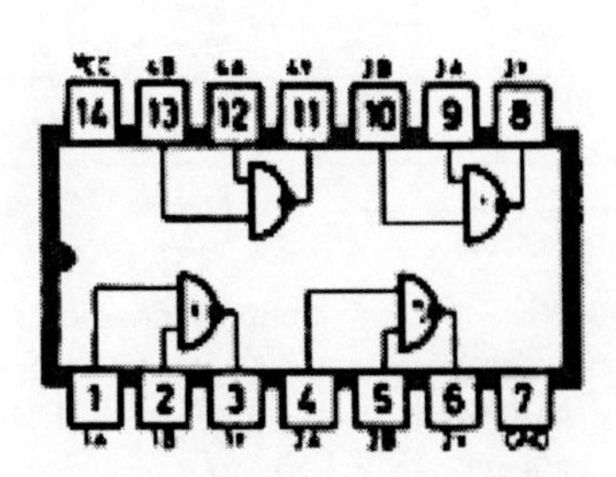

The 7400 TTL logic circuit

Around 1971 the first microprocessor was developed by Intel, in which most of the computing circuitry was contained within a single chip. The response of people in the mainframe computer business shrugged this development off with "what's it good for"? In a microprocessor, like the tab on a tabulating bar, each program step still performs the same function every time it is referenced. More elaborate branch and compare functions allow conditional processing of one or another set of functions. Not much has changed in this model. Clock speed, the rate by which program steps are executed, has increased and processors have become more and more complex, having been optimized to do as much as possible in the shortest possible time.

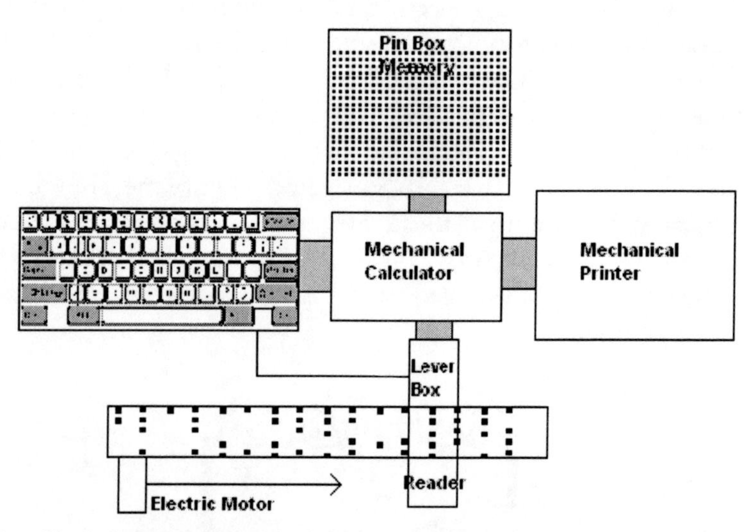

The old tabulating machine block diagram.Mechanical counters consisted of gears that were assigned binary values of 1,2,4 or 8. Memory consisted of a pin box in which pins had binary values. Levers 'read' the tabulating bar and enabled the mechanical keyboard when input was required and transferred the pinbox memory to the accumulator, etc.

In a 'regular' von Neumann computer, no matter of which generation, a program that is stored in memory is executed on a central processing unit (CPU). All data and program instructions are stored in memory, expressed as sequences of ones and zeros. The program acts on data that exists in memory, or is requested from the operator through an input peripheral. The program and data that is in memory are initially loaded from disk or another form of non-volatile memory. The processed data is eventually sent to an output peripheral, such as a file on a disk drive or a printer.

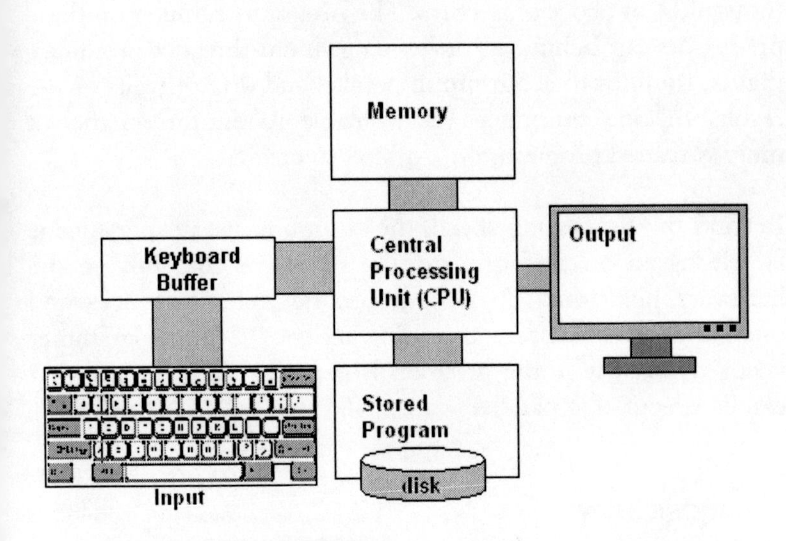

The block diagram of a 'von Neumann' computer

Over the years, many advances have been made to optimize this model. Multiple processing cores, optimized data pipelines, fast cache memory and multiple ALU (Arithmetic Logic Units) have been added to perform some tasks in parallel and to eliminate waiting cycles. Processing speed and the number of transistors in a processing unit has increased according to Moore's law, which states that the number of transistors on a single chip doubles every couple of years. The technology is now approaching its physical limits. The physical properties of silicon change dramatically beyond a certain level of miniaturization. In silicon, smaller means faster. Processor clock speeds have gone up to 4 GHz, which is in the microwave region of the frequency spectrum. The only way to increase processing speed in silicon chips is to use multiple parallel processors. All the processor manufacturers are now offering multi-core processors and this is a trend that will continue. Future processors will have hundreds, if not thousands, of processing cores. The programs running on these processors lag behind in developing 'multi-threaded' routines, that is, routines that can run in parallel and do not wait for the results of one routine to be available to another. Efficient multi-threaded programs are difficult to create.

Beyond the processing speed, the computer user experience is largely based on the software, the programs that run on the hardware platform. All modern programming languages and operating systems use a threading model. If the programmer keeps this in mind, the program is compiled into threads that can be executed in parallel.

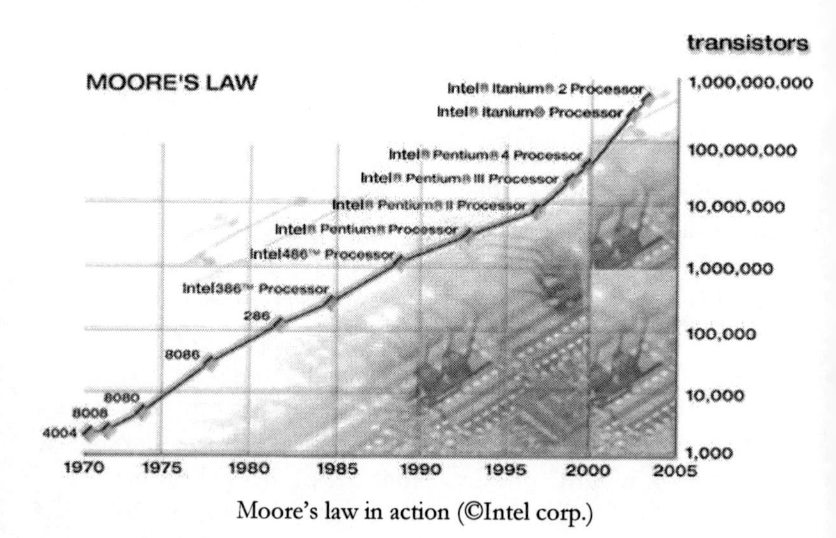

Moore's law in action (©Intel corp.)

The clock speed of a processor has long been a marketing standard. The clock signal determines when and how quickly the internal state of the CPU changes. Clock speeds become irrelevant as the number of processing cores increases. This will change dramatically with the future introduction of clock-less processors; devices in which the throughput is not determined by a clock signal but by the rate at which data is applied. The number of processing cores will be the new marketing standard. Microprocessors, Signal processorsand Graphics processors are variations on the Von Neumann computer. They are miracles of modern engineering and well suited to perform fast mathematical calculation. Even though many advances have been made that have increased the processing speed of these CPUs, the fundamental architecture is unchanged since von Neumann determined the architecture of the stored program computer in the 1940's. Its method of fetching program instructions, executing them and storing data in memory is still reminiscent of its tabulating machine heritage.

Artificial Intelligence programs that execute on such machines are control programs that interpret incoming data for the largest part through sequences of logic comparison statements, sequentially comparing input data against stored lists. Given a defined sequence of stimuli, a control program will always produce the same result for the same set of stimuli, e.g. the output produced by a range of stimuli is a predefined set of results. The decision tree is used to control the process. Given an unknown stimulus the decision tree will fail to produce a result and the process will fail. The result could be random or nil, resulting in either random behavior or no behavior at all. Imagine a military robot that is controlled by such a process. It will function perfectly well 99.99% of the time, until something unexpected happens. Much of the fear of Artificial Intelligence systems(and portrayed in many Hollywood movies) is based on this unpredictability of behavior. There are also practical examples where unpredictability of behavior or programming errors in critical applications have caused loss of life. In 1977 an Air New Zealand DC10 jumbo jet with 257 people on board crashed into a mountain slope in Antarctica. The computer was flying the plane along a set of coordinates to a predetermined flight plan. Two of the coordinates in the flight plan were wrong. Computers don't question what they are doing, they just run programs. Programmers are human and make mistakes. Such mistakes have caused rockets to disintegrate shortly after launch, caused car engines to stall and airbags to deploy at random.

Artificial Neural Networks

An Artificial Neural Network (ANN) is based on a 1950's model of a neural cell. It uses static 'weights' that are assigned to inputs. All the weight values for which an active input exists are summed, and an output value is produced if their combined value is higher than a threshold value.. Within a standard computer, the ANN program creates layers. The first layer is the sensory layer. Then there are one or more hidden layers, and finally an output layer. A known pattern on the input of the matrix produces an output. In an ANN the neural cells in each layer exist out of a small decision tree that determines if the input parameters of the 'cell' have been met. Real neurons are not such static pattern matching devices. The output of a real neuron is an integration of spatial and temporal distributed pulse trains. ANN networks do not learn autonomously. Learning takes place by manually modifying the synaptic values in the hidden layers and the output layer. The neural network is trained by presenting it with a known input then manually modifying the weights in each of these layers until a desired response is obtained. Neural networks presented with an inexact set of stimuli can be trained to produce sensible outputs. Structurally, neural networks are suitable to parallel processing and are straightforward to implement in hardware. The network cannot deal with complex, unknown problems because the network is static, e.g. no learning takes place after the initial, manual training. Neural networks are capable of recognizing patterns where the pattern that is presented to the network is incomplete or distorted, and have been successfully used in areas such as handwritten-character recognition and writer identification where the input varies or is inexact. Hopfield networks are an improvement on static neural networks. Hopfield networks work with an 'energy' value. They learn by minimizing the 'energy' level in the node. Unfortunately they are very slow to train, and still behave much like a static neural network, and not at all as an interactive information carrier.

Genetic Algorithms

The Genetic Algorithm (GA), developed by John Holland in the 1970's, is a programming method that claims to copy the process of 'the survival of the fittest' in natural evolution. Genetic Algorithms are used in pattern recognition, computer games and in optimizing sorting algorithms. Genetic Algorithms involve the 'mating' of successful bit-patterns, whereby sections of the bit-patterns of both parents are inherited. A random element is introduced by flipping one or more bits in the inherited bit-pattern. The offspring is tested for fitness against a known data set. If successful, the offspring is maintained and will become a parent. If not, the offspring is destroyed. This produces a highly optimized set of patterns within a known scope, starting with a random set. The most difficult and vital part of the genetic algorithm is the fitness test. The fitness algorithm must preserve the best performing descendants. The population is constantly culled to the original number by eliminating the descendants that performed worst.

Used in computer games for instance, the fitness test is determined by winning or losing the game. The more games are won, the better the quality of the algorithm.

Example of a Computer Control System

Control and measurement systems are everywhere. The input can come from virtually any process, such as the digitized medical information from an Electro-Cardiogram machine, digitized human speech for speech impediment studies or the minuscule voltages generated by materials when they are heated. To illustrate how this works, consider the simple temperature controller in the figure across the page. A Thermocouple that is labeled **T** generates a tiny voltage that is proportional to the temperature of some material **M**. This voltage signal is amplified in **AMP** and fed into an Analog to Digital converter. The result is a 13-bit (2^{13} = 8192 steps) binary value that is read by the microcontroller under program control. If the temperature range is 0 to 1,000 °C then each step in the digital data is 1000/8192 = 0.122 degree. The thermocouple response curve is logarithmic and the program code has to compensate for this. The program switches on the heater coil through the Heater Control switch. In this case there are four control lines, so there is a maximum of 16 levels. The Microcontroller reads the set point temperature from switch 'Set Point'. This is a 16-position switch with 4 binary outputs, so each setting is about 60 degrees. The program keeps the heater on until the set point is reached. The temperature is displayed on a 3-digit, 7-segment display that is controlled by the digital output of the microcontroller.

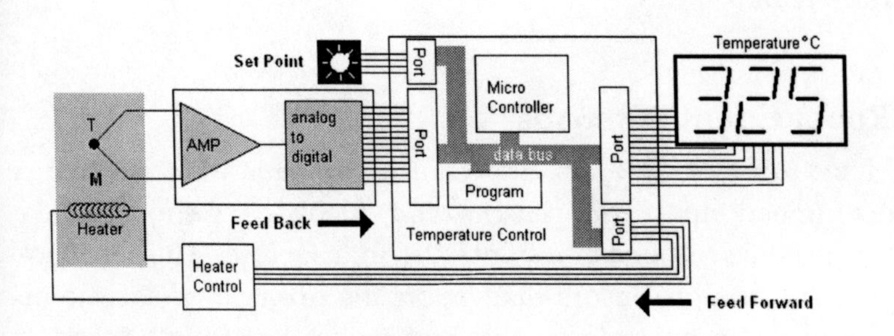

The 'perception' of the temperature controller is limited to a single 13-bit binary value and a 4-bit set point value. The temperature is controlled by the control program and the programmer has made certain assumptions. The control algorithm is often a PID loop (proportional–integral–derivative), whereby the heater power is calculated proportional to the measured temperature (feedback). The programmer has determined the binary switch position that represents a preset temperature. The programmer also determines, through a calibration process, the meaning of the binary output of the Analog to Digital converter at different temperatures. The control of the heater coil is the feed forward path. It directly controls the temperature. The value returned from the process is the feedback. It tells the program something about what is going on in the process. All control processors, no matter how complex, are variations on this theme. Large networked systems consist of many small controllers that have a network interface. The setpoint is often controlled from a central computer rather than a rotary switch. Control systems that use feedback are not new. They have been used for a long time before computers existed. An example of a mechanical feedback system is the centrifugal governor on a steam engine, which uses weights pushed out by centrifugal force to control the engine's speed.

Robot Control Systems

Most robotics systems are control systems that receive a determined input from sensors, and calculate a response. Take for instance a robot arm controller in a factory. The feedback path consists of sensors that report the position of each joint. The program controls motors that move the arm in X and Y directions in small steps, and another motor to control the elevation. In this example, four bits are used to control the stepper motors, resulting in 16 possible positions. Stepper motors do not rotate constantly, but performs small steps that correspond to a binary value. A 16-step motor performs one 360 degree rotation when the binary codes 0000 counting up to 1111 are applied. This is a very simple robot arm. It uses the same controller board as the temperature control on the previous page, but with a different program stored in its memory. Industrial robots generally have many more joints with more precise control and feedback sensors. The sensors are circular encoders that report position as a binary number. In this case the binary values range from 0 to 15, providing 16 different positions. The position value is input to a microcontroller under program control. The program compares the set point to the current position, and the difference is used to control the stepper motors until the difference is zero.

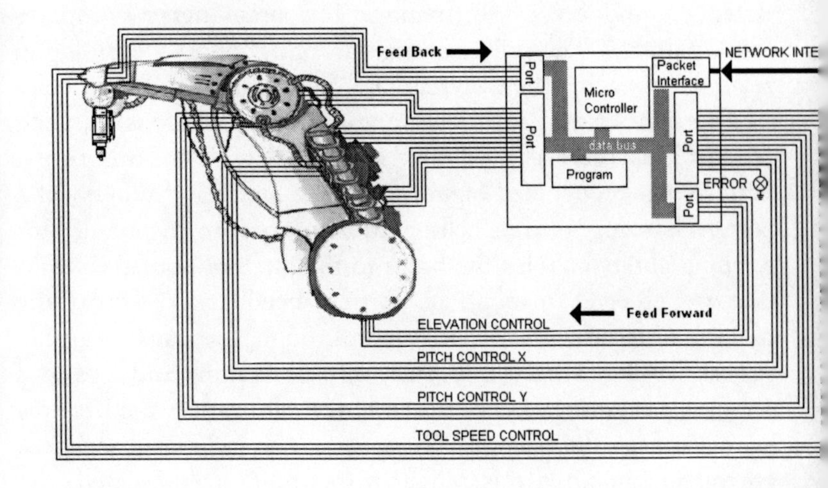

Humans Have Feedback Too

Human sensory perception is nothing like the sensors that are used in control systems, nor are our limbs controlled by some DNA encoded program. Our movements are learned and stored in synapses that are updated constantly. The function of feedback in the brain is twofold – to learn or confirm stored memory and to alter or produce the pulse trains that control the body. The brain's motor cortex is trained early in life. All sensory perception is communicated in the shape of timed pulse trains. These pulse trains are digital in nature. There are no Analog to Digital conversions, and no variables. In the eye for instance, 180 million light sensitive nerve cells are concentrated in the fovea. The image is projected by the lens onto the fovea at the back of the eyeball. As light hits each cell, it switches, causing a pulse or pulse train. This 'translates' the entire picture into pulse trains that are relevant in spatial relation to each other and in timing. These pulse trains are further processed in the eye to provide the brain with edge detection and color information. The brain never compares these patterns. The pulses contain information in the timing in relation to each other, and in the distribution of active cells. The feedback paths in the brain are massive. There is as much feedback as there is feed forward. Feedback confirms that a pattern was recognized higher up in the hierarchy. Whenever a pattern is recognized its cell-based memory is re-enforced. This learning ability enables the brain to build a model of the world that it perceives through its senses. Feedback confirms the pattern and errors are corrected through the same learning process. When a baby is born it moves its arms and legs in a seemingly random way. But each movement causes the brain to learn. At first, the changes that are made to individual synapses are huge, but after this initial 'formatting' is completed, the changes become gradually more subtle. Learning enables this memory system to remember how to move a limb, rather than to control its movement. The entire brain functions according to the same algorithm. We never stop learning. We learn how to stand up and walk as an infant, but we add new information to the motor cortex later in life, maybe at age 50 when we learn to dance.

Sensory Information is Relational

The pattern that is input to the matrix is always generated by a large number of parallel sensors in relationship to one another. The eyes, the tongue, the nose, the ears and the skin each have millions of specialized sensors. If we close our eyes and we are handed an object, we can feel its shape and its surface features. We form a picture in our mind. This is because our large parallel sensor 'skin' array outputs pulses that have a time function. The spatial pattern is formed by nerve cells that are adjacent, while the timing of the pulses holds temporal information so that we can build up a picture in our mind of what we are holding. The same is true for hearing and vision. We can locate a sound by the temporal differences that each ear perceives. We see depth in a picture because of the differences in perception between the eyes. We can illustrate how this works by using a simple example. In the following diagram the plates respond to the touch of a finger. The sensors in the diagram are simple on/off sensors. They switch on in contact with the skin. As we move a finger over the sensor array, a set of sequential pulses is generated, as is shown in the diagram. The pulses are shifted in time and spatial orientation. This is similar to what happens in our skin. The sensory nerve cells in our skin connect in parallel to a matrix of neurons, the somatosensory cortex, in our brain. This is how the brain forms a 'picture' of the object in our hands.

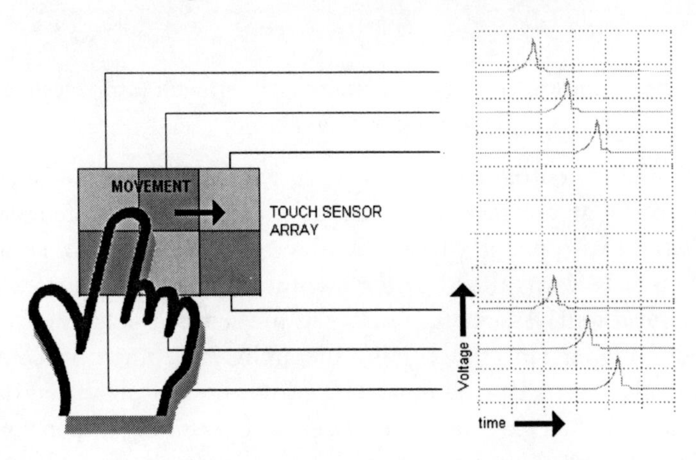

Demonstration of Spatial/Temporal pulses

The input to the neural matrix is an array of sensors that generate spatiotemporal pulses. In the emulated brain, the sensors can be pressure switches, light sensors, heat sensors, artificial cochlea or even sensors for which we humans have no equivalent. These sensors must be arranged in an array to get meaningful patterns for cognitive applications. When the device is used to move a robot arm the movement will be random and uncoordinated at first. Through feedback from position sensors on the arm the device learns how to move the arm in a coordinated way.

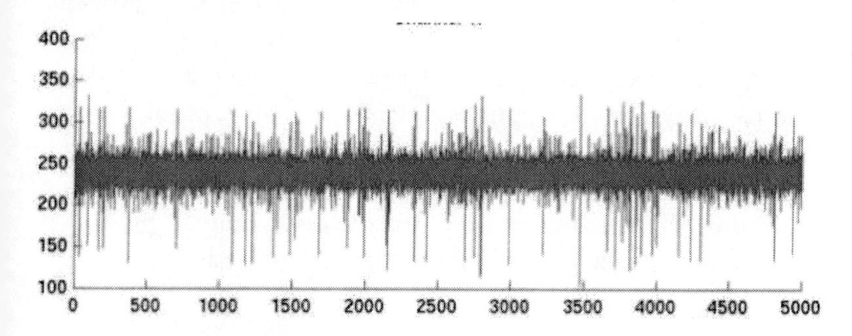

A representation of audio frequencies in speech (amplitude and frequency over time)

A synthetic neuroanatomy processor will learn temporal/spatial patterns that are repeated without the need for labeled data. During the training phase the device is exposed to simple patterns. It learns those patterns without a need to know what they represent. Association between the senses is accomplished much higher in the brain hierarchy. A microprocessor controller needs labeled data. The conversion of unlabeled data to labeled data can be accomplished by reading the neuron registers during device training. Neuron activation registers and synapse registers located higher in the hierarchy will be significant, and can be labeled by associating the stored values with input patterns during training. It will not be necessary to label data in systems where no microprocessor is used. Multiple Synthetic Neuro-Anatomyprocessors all use unlabeled data.

The Synthetic Neuro-Anatomy has an access port for a microprocessor. The signals that are present at the outputs of the device are not useful as inputs to a microprocessor. The outputs are intended for connection to other Synthetic Neuro-Anatomy processors to build larger systems. The access port maps all the registers of the synthetic neurons and the synapses as memory. Relations between values in memory and real-world events are established during the device training. An unlabeled value is meaningless to a microprocessor, but all values in the synthetic neuron matrix are unlabeled until they are identified and labeled by a programmer.

Chapter 14

Tools for Brain Research

"If we knew what we were doing, it wouldn't be called research, would it?"

Albert Einstein

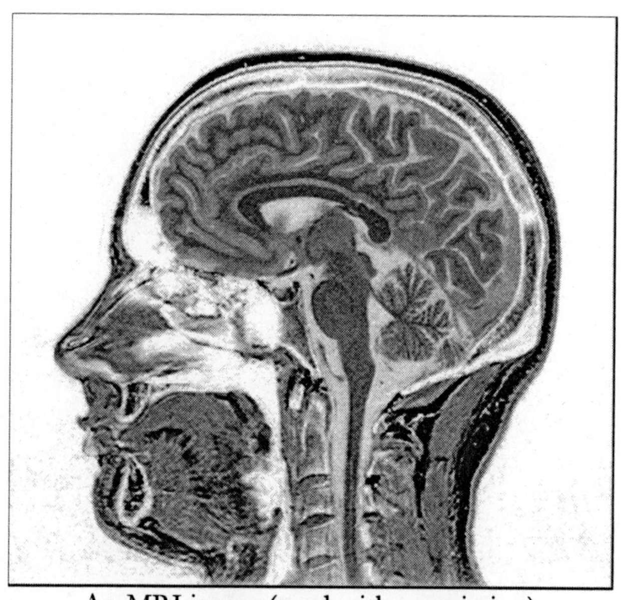

An MRI image (used with permission)

Tools for brain research have improved greatly in recent years. We can now 'see' a living brain in operation, something that was unthinkable a few years ago. Improvements and new discoveries are made constantly but we still have some way to go before we have a complete map of a living person's brain. Before MRI and PET scans, the only technique to view the brain's structure was to 'cut and slice' the brain. Each thin slice was examined under a microscope and documented. This is a long, tedious process and it is not practical to map 86 billion neurons in this way. This could obviously not be done on a living person. After death a brain goes through immediate changes. Its 'software' disappears just like that long unsaved document disappeared when your computer switched off in a blackout. Examining a 'dead' brain cannot give us enough information to build a synthetic brain. The connections will be intact, but the content of synapses is lost.

The computational primitives of the brain are not just neurons, but also the synapses and the surrounding glia. Each synapse is just 1 millionth of a meter in diameter. This small size and the fact that there are about 100 trillion of them make it impossible to determine their individual state using existing technologies. Synapses contain vesicles that enclose a particular neurotransmitter. Each neurotransmitter has different properties. There are hundreds of different neurotransmitters. All the synapses connected to the same neuron contain the same neurotransmitter, but the size and quantity of vesicles vary. To model a living brain, we would need to 'read' the content of each synapse at the molecular level, at a sub-nanometer scale. Mapping 100 trillion synapses is at best a challenging task.

Magnetic Resonance Imaging

MRI (Magnetic Resonance Imaging) was invented in 1977 by Dr. Raymond Damadian. The machine has a long tube, about 60 cm in diameter in whichthe patient is immobilized during the scan. The tube, known as the bore, consists of a very strong magnet. The magnetic field inside this magnet is about 30,000 times stronger than the earth's magnetic field. This equates to a magnetic force of 1.5 Tesla (15000 Gauss). The magnetic field is generated by a superconducting electro-magnet that is cooled using liquid helium at 450 degrees below zero. A number of gradient magnets 'shape' the magnetic field to focus on different parts of the patient's body. Most hydrogen atoms in the patient's body line up to the magnetic field, which runs straight down the center of the tube. Some atoms do not align until a strong, high frequency pulse, resonant with hydrogen atoms, is applied. The misaligned atoms absorb energy and turn. When the high frequency pulse is removed, the atoms turn back, emitting energy. It is this energy that is mapped and used to create images. Each element has its own specific resonant frequency. The density of hydrogen or oxygen can be mapped by using different frequencies. New MRI machines use extremely intense magnetic fields, up to 21 Tesla. These high-Tesla MRI machines have a better resolution, and are faster than regular 1.5 Tesla machines, but the long-term effects of exposing humans to strong magnetic fields are uncertain.

Electroencephalogram

To identify the strength of the synapses and 'read' the information that is stored in the brain we need tools that can accurately and rapidly scan large numbers of synapses at the molecular level. Those tools do not exist as yet. Neurons integrate the output of many thousands of synapses. Neurons come in different sizes, ranging from 4 micrometers to about 100 micrometers. Glial cells form a support framework for the neurons, but they also participate in processing and in the process of synaptic plasticity. This process is not well understood. Glia are thought to synchronize groups of neurons. It is this synchronizing effect that gives rise to 'brainwaves' such as Alpha, Beta and Gamma waves seen in EEG and MEG scans. EEG uses relatively large sensors attached to the scalp with a conductive jelly. MEG is similar to EEG, but is a more accurate technique. MEG does not have a physical contact with the subject, but uses superconductive magnets and coils to pick up the minute magnetic field generated by the potentials in the brain.

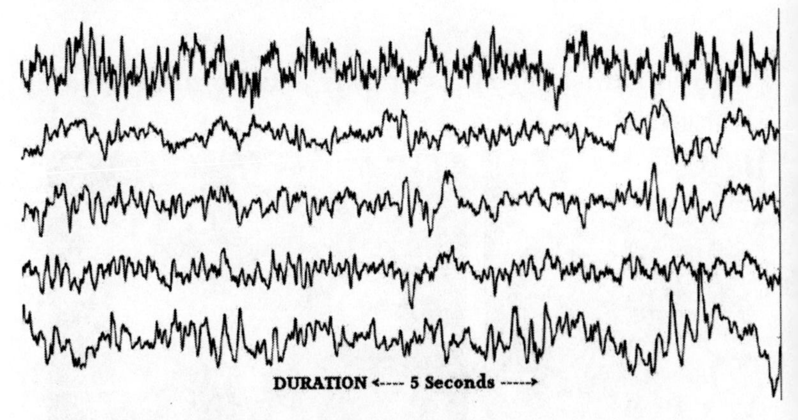

Typical EEG graph output, showing 5 channels recorded from electrodes placed on the scalp.

There are many millions of neurons under each sensor. The dense skull forms a barrier that has a blurring effect. These scans give us a general picture of what the brain is doing. It is like attaching sensors that measure the electrical field on the outside of a computer to determine what program the machine is running. With sensor loops and a multi-channel oscilloscope it is possible to determine if the machine is idling or if it is running a program that involves intense processing. Depending on the location of the sensors it is also possible to pick up the rate at which Random Access Memory is refreshed. The results are very similar, although at a much higher frequency, to what is seen in an EEG graph. The temporal resolution of these systems is excellent, about 1 millisecond, but the drawback is that the response is the result of many neurons firing. Interesting results can be obtained when the EEG is combined with activity. If the brain is stimulated, it responds with a characteristic wave pattern. Analysis of these wave patterns and their distribution across the scalp gives an indication of the network dynamics of the brain.

Positron Emission Tomography and Computer Aided Tomography

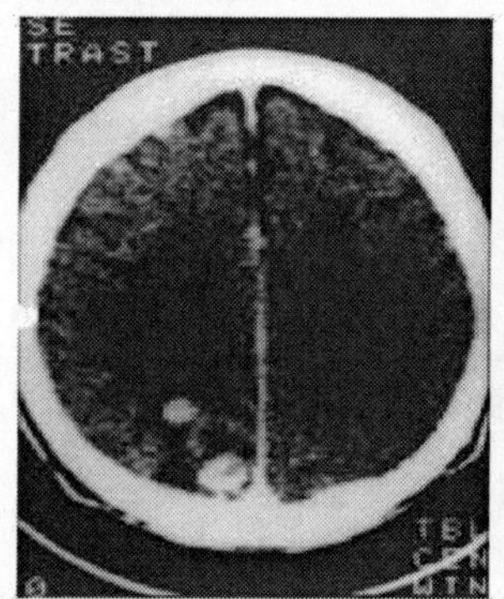

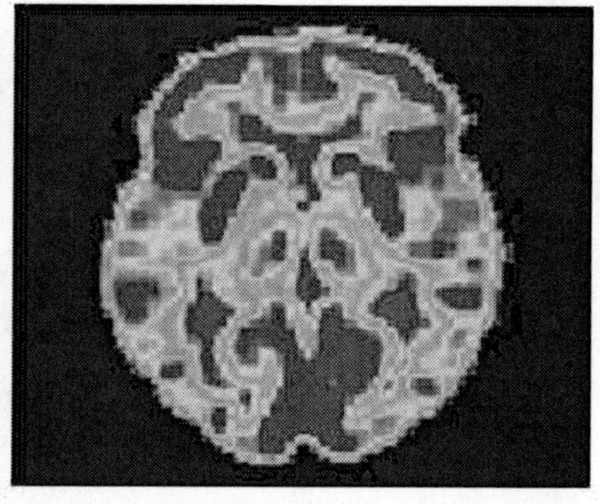

CT image of a brain tumor (left) and a PET image of a healthy human brain (source: public domain US National Institute on Aging)

In PET/CAT (Positron Emission Tomography/ Computer Aided Tomography), a radioactive agent is injected into a vein and carried to the brain in the bloodstream. Because of this radioactive agent, there is substantial risk to pregnant or breastfeeding mothers. PET scans are not recommended for young people. The CAT scanner creates a composite picture of the brain and the distribution of the radioactive agent.2-DG Imaging is used in conjunction with PET/CT scanning. 2-DG is a modified glucose (sugar) containing fluorine-18, which emits Gamma radiation. 2-DG is taken up by the glucose transporters of the cells as if it were normal glucose. The substance is slightly toxic, and marks areas of high glucose uptake. Cells with a higher glucose uptake such as cancer cells will light up by gamma emission, which is imaged in a camera. Because the glucose is modified, it is not converted into energy and therefore slows cell growth.

Tomography refers to any imaging method that shows 'slices' of the body. Traditional tomography involved the removal of the brain after death. The tissue was then hardened with formaldehyde, after which the brain was sliced like a ham. Each slice was studied under a microscope and photographed. Modern radio-tomography methods are not quite that destructive. The traditional cut and slice method is still used to examine the brains of laboratory animals and in human post-mortem examinations.

Single-photon Emission Computed TomographySPECT

Single-photon emission computed tomography is a nuclear medicine technology that detects gamma rays, emitted from a nuclear isotope that is injected into the patient.

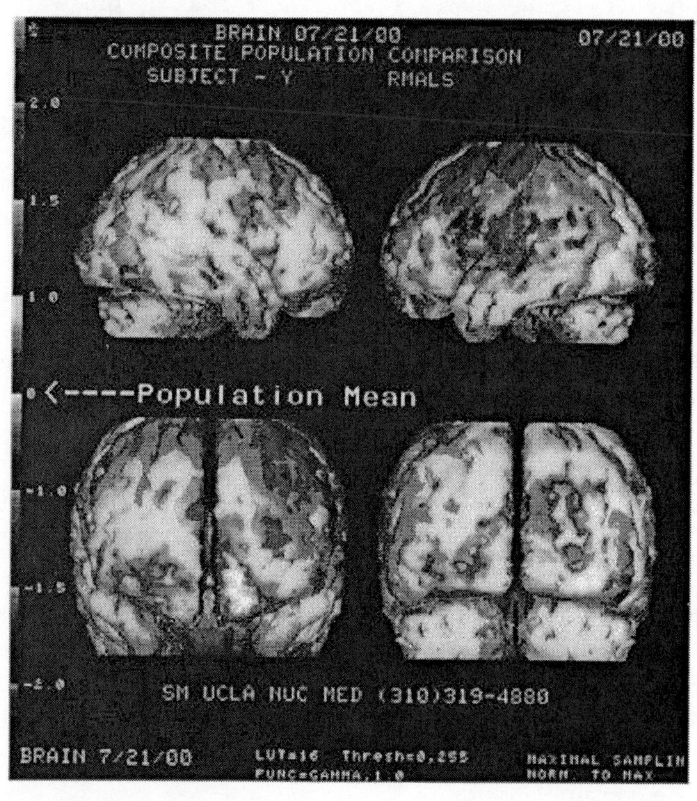

3D SPECT images © UCLA Nuclear Medicine

Summary of Brain Imaging and Research Tools

Research Method	*Description*

EEG and MEG	Electro-encephalogram and Magneto-encephalogram.
2-DG	2-Deoxy-D-glucose. A marker substance
Cellular Imaging	Optical light microscopy
Patch clamp	A glass micro-pipette is placed over a 'patch' on the cell. The patch area is one micrometer. This area covers a single channel. A silver wire is inserted to record current flow.
Field potentials	By using an extra-cellular microelectrode the electrical potential is measured in surrounding tissue.
Optogenetics	An emerging technology that uses genetics to change the optical properties of neurotransmitter proteins (glow-in-the-dark mice).
Microstimulation	Insertion of a small electrode into the brain, and observing the results when a minute current is injected into neurons.
Optical imaging	Measures the diffusion of near-infrared light in brain tissue.
Tomography	Traditional imaging method that 'slices' the brain into thin sections and uses microscopy

	to examine each sample.
MRI and fMRI	Maps the energy that is released by excited hydrogen or oxygen atoms (fMRI). The atoms are excited by a high frequency pulse.
PET Imaging	Positron emission tomography is a 2D imaging technique using an injected nuclear isotope.
SPECT	Single-photon emission computed tomography. 3D imaging technique that is using a gamma ray emitting nuclear isotope, injected into the patient.

Chapter 15.

Smart Computers, Artificial Intelligence So Far

"I've finally learned what 'upward compatible' means. It means we get to keep all our old mistakes."

Dennie van Tassel, Author

Photograph By Andriusval [Public domain]

Browsing the New York Times archives is an interesting journey into the past, giving us a view of the contemporaries at the dawn of great inventions. The misconception that computers are some sort of brain machine has existed since the dawn of the era. Journalist failed to understand the concepts of logic and programming, and depicted these machines as 'Brains'. This spoke to the imagination but also scared a lot of people. The result was 'pulp' science fiction in which robots were the master brains that ruled a reign of terror over people. People felt inferior compared to the marvels that these machines were supposedly able of. Scary stories sell better than then truth that these giant machines were little more than programmable calculators. They underestimated the cognitive aspects of the human brain. Neither did the scientists at the time understand the limitations of their machines, nor did they fully appreciate the cognitive abilities of the human brain. To them, cognition was just the next step in a long line of successes. At the time the newsstands were flooded with fictive tales about robots, portrayed mostly as murderous machines and heroes with superhuman capabilities. More than a few of those were stimulated by the 'giant brain' newspaper stories, and the collective feelings of inadequacy that they inspired. If there were superhuman machines, then there had to be superhuman heroes. It is easy to get lost in the online newspaper archives, and tempting to look up great names such as Thomas Edison, the Wright brothers, and Thomas Watson. Edison was considered foolish by his contemporaries for his efforts to develop electric light while gaslight worked so perfectly well. At $3 a light globe it was considered far too expensive, much more so than a gas sock. A gas sock is the fragile filament that is put over a gas flame. It glows and turns heat into light. Further, the critics said it would never work. We would still be using gas light today if Edison had listened to any of these people, who appeared to make perfect sense. Then there is the famous quote from Thomas Watson, chairman of IBM, who said in 1943: "I think there is a world market for maybe five computers." Or the engineer at the Advanced Computing division of IBM who said in 1968, "But what ... is it good for?" when commenting on the microprocessor.

No Intelligent Machines Yet

Movies like 'StarWars' and '2001: A Space Odyssey' have primed the public's mind to the possibility of intelligent machines like C-3PO and HAL. More than thirty years later the reality is very different. To put it bluntly, an intelligent, aware and thinking machine does not exists, it is not even a remote possibility at this time. Existing industrial and experimental robots are controlled by a program and have neither awareness nor intelligence. Programmed digital computers appear to be a blind alley as far as intelligent machines are concerned. They are 'cause and effect' machines. A whole new technology is required to build an intelligent machine, a technology borrows heavily from the structure of the brain. The brain is the only machine that is capable of acquiring knowledge and performing cognition and reasoning tasks. One possible way to build a thinking machine is to accurately emulate the entire brain and all its methods. Efforts to do this have so far concentrated on creating complex simulation programs on fast computers using millions of processing cores. Computer emulation of the brain requires not only huge computer resources but also a large team of programmers and scientists, and many Megawatts. What can we really expect from these efforts, will a programmed simulation of the brain show intelligence? Will it have the capability to learn? If the model shows neither intelligence nor is capable of learning then the question must be raised if it is a valid brain model. The Blue Brain project focus is to emulate the entire brain at a precise molecular level in software. The aim of the project is not to build an intelligent machine, but to understand how the brain works. Because the brain is an intelligent entity, it is difficult to see how the two can be separated. It will be interesting to see if the result will be a thinking machine, and if this machine will be capable of learning. To learn in the real world (as opposed to a simulated world), the machine will need to work in 'real time' whereby 1 second of emulation time is 1 second of brain time. The machine will, like the brain, need to be synchronized to the world.

Machine Intelligence

The path to Artificial Intelligence is scattered with failures. It is impossible to develop an intelligent machine without an understanding how the brain develops from infancy to adulthood. The brain is a machine that acquires knowledge and evolves intelligence, but its methods have been obscured by the complexity of its massive number of components and interconnections.

In his 1950 paper 'Computing machinery and Intelligence", Alan Turing presented a very limited view of intelligence. He defined a machine as intelligent when its responses in a conversation are indistinguishable from a human being. This is known as the 'Turing Test'. It is performed by conducting a text-based conversation, in which the participants do not see each other. Two participants are human, one is a computer program.

The responses are selected from a database, through a methodical analysis of the human reply that is entered on a keyboard. The databases and the selection criteria are created by a very human programmer, and it is his or her intelligence that is displayed by the machine. Really, it is two humans that we converse with, except that one has pre-recorded all his responses and written a program that selects one of those responses. Humans do not have a database of pre-recorded responses. We make them up as we go. Conversing with one of these programs can be funny, because often they fail to analyze the user's responses correctly and come up with random remarks. It is interesting to see what the machine's responses are when we give if a sentence that consists of random phrases.

Natural Language Processing

ELIZA is one of the oldest of the original chatterbot programs. It is a free open source program that tries to carry on a conversation along Turing's line of thinking. It uses a process known as 'Natural Language Processing' to analyze inputs. NLP is a process that searches a database of words and looks for matches. The program selects a response from a database of pre-recorded responses. The selection depends on the matches it finds. If no match is found the program selects a global response. This is not intelligence. It is simple selection. There is neither understanding nor thought. How often do you sort through a list of phrases in your mind to understand another person? Gonzales Cenelia, who translated the original Pascal code, states in the source code comments that, "The general view about this subject is that it would take many decades before any computer can begin to understand "natural language" just as humans do." The original Pascal version of ELIZA was created in 1987.

Elbot is an on-line chatterbot that works by the same principles. It can be fun to play with these things and see the nonsense they come up with when there is a high level of ambiguity in the human responses. An entertaining experiment was done by students at Cornell Creative Machines Laboratory. They took two instances of CleverBot, and got them talking to each other[44]. Keeping in mind that all the responses are programmed by humans, and that the same program was in effect talking to itself, the result is still captivating to watch. Some hype was created and the YouTube clip was a success.

UK philosopher A.C. Grayling states about the Turing Test;

> *"The test is misguided. Everyone thinks it's you pitting yourself against a computer program and a human, but it's you pitting yourself against a human and a computer programmer - i.e. two humans."*

Turing and other A.I. researchers never made any mention of learning. The processes of the brain that Jeff Hawkins mentioned in 'On Intelligence', such as Prediction, Association and Invariant recognition[45] cannot function before the neural matrix is filled with information through learning. Learning is a fundamental function of the brain. While its basic layout is hard wired by DNA, by learning the brain formats its processing centers for speech, vision, taste, behavior, and social interaction.

---

[44] Hod Lipson, Jason Yosinsky, and Igor Labutov at Cornell Creative Machines Laboratory

[45] Jeff Hawkins and Sandra Blakeslee, On Intelligence ISBN 0-8050-7456-2

Computer Intelligence and IQ Tests

Let us consider what it would take for a computer program to pass an I.Q. test. Such tests include sequences of numbers, figure transpositions and word combinations to complete. To make things easier for the programmer, we assume that the test questions are in an electronic document format so that the machine does not need to have visual recognition capabilities. Furthermore we assume that all the test questions are known to the programmer, that the test comprises a sequence of prime numbers 2, 3, 5, 7, 11, 13 (numbers that can be divided only by 1 and themselves) and that the next number in the sequence is what is required. The program will need to evaluate the number sequence to determine that this particular test contains a series of prime numbers. Then the program can calculate the next number (17). A similar analysis is possible for word combinations and scrambled letter combinations. Shapes are more difficult to analyze, but assuming that they are simple shapes our programmer may find a way to interpret the data. We now have a 'smart' program that can pass this one specific I.Q test. It is obvious that the program will fail if we change the questions. It is not too difficult to change the program so that the questions can be in a different sequence, as long as the questions include the calculation of the next prime number, recognition of a shape and word combinations.

When we introduce a higher level of variation in the tests, the program would fail on those questions that were not along the same lines as the original set. The programmer would have to design new code to cope with the new questions. The processing time and complexity of a program increases linear to the variation in the test questions. Hence, an open set of I.Q. test questions increases the complexity and processing time by a factor approaching infinity. The only solution is to make compromises that limit the test questions to a defined set. The environment in which the machine operates has to be simplified to suit the program.

This simple example illustrates the limitations of 'Artificial Intelligence' controlled by program code. The programmer will need to know all possible situations that the machine can encounter. The real world presents an infinite set of situations. To get something working, a programmer will need to make a number of assumptions. This will mean that the program is unable to meet all possible situations. We see occurrences of this problem in speech recognition, face recognition, robot control and trend forecasting programs. These programs consist of long lists of features and routines that search the input for matching patterns and calculate probabilities. The real world does not conform to a limited set of patterns that can be matched against a database. Therefore the programs that operate by these principles work to only a limited extent.

Chapter 16

The Ethics of Artificial Intelligence

"A year spent in artificial intelligence is enough to make one believe in God."

ALAN PERLIS, Artificial Intelligence: A Modern Approach

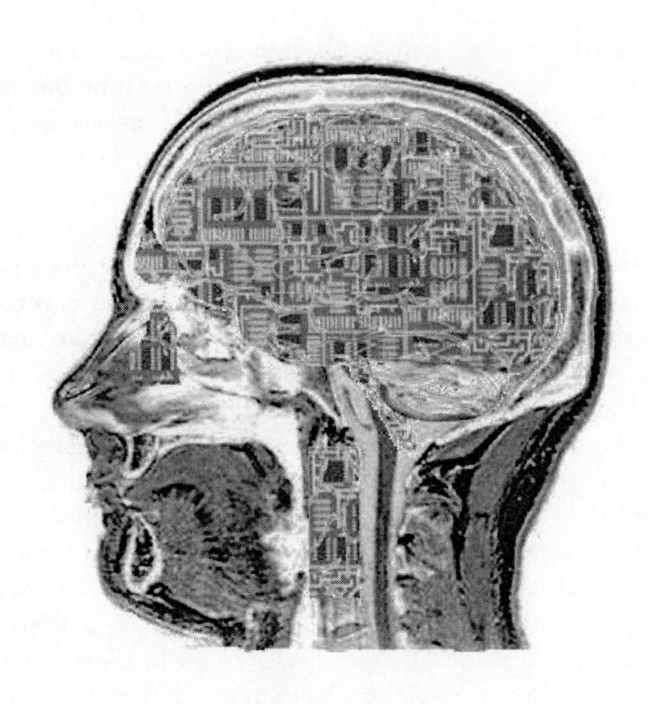

A.I. ethics is already a diverse area of study whilst the actual machines are still some time in the future. Books like 'Moral Machines' by Wendell Wallach and Colin Allen, and 'How to Survive a Robot Uprising' by Daniel Wilson speculate on what these machines will be capable of, and how humans will interact with them. Ethics determine the obligations of society towards intelligent machines, and their obligations towards us humans. It identifies the principles that rule behavior. The key to determining the rights of intelligent machines is whether these machines are sentient. So far only the 'programmed dolls' - cause and effect machines have been considered, and they are not sentient. How will we determine if a machine is sentient? Animal rights people determine that something is sentient if it can experience pain. That is too narrow. Sentience is closely associated with awareness, having conscious thought and intelligence. We have not yet succeeded in creating intelligence the size of a cockroach, so all the suppositions in books, Internet articles and magazines about 'super-intelligences' is foolishness. A lot of words like 'probably' and 'presumably' are used in those articles – everyone expects these things to evolve, but they never do and never will, unless they have the ability to learn. New questions need to be raised with the event of an emerging artificial intelligence within a synthetic neuro-anatomy processor. What will be the status of such a machine in our society, and how is society going to respond? In what way are their legal rights different from ours? The emergence of intelligent machines will be a gradual process, not a sudden event. This gives us time to develop the right ethics, as the machines get more elaborate. Machines that have only a few million neurons, and therefore a low ability to develop intelligence, will appear first. As their brains continue to grow with the introduction of larger devices, new intellectual abilities will be added over time, just like each generation of computers has added new capabilities over previous models.

Integration in Society

Sentient robots will not 'grow up' and evolve with a compassionate, helpful, good-willed and gentle nature all by themselves. They will need to be trained. With any new technology come issues of integration, legal matters, security, and social acceptance. The problem of machines turning against humans is generally known as the 'Frankenstein Syndrome'. 'Terminator' depicts a future in which the machines and humans are at war, with most of the human race wiped out. In the movie 'RoboCop' the machine goes mad and kills indiscriminately on a grand scale. The list of movies about mad or dangerous robots is extensive. In 'I Robot' an evil genius creates a generation of uninhibited robots. Hence, the robots are going to be used to wipe out their human enemies. Most Hollywood robots have very poor ethics (this makes sense somehow). The 'Star Wars' movies are an exception in that the robots seem well integrated into society as useful tools. This model appears more realistic. As with any new technology, there will be a significant impact on society, similar in scope to the industrial revolution and the more recent computer revolution. The largely home-based industries of the 18th century all but disappeared when steam engines were introduced. Factories were established which could produce goods much more efficiently and cheaply. In our lifetime we have seen the shift from manual labor to an automated, computer-based society and from factories with hundreds of workers to factories with industrial robots and a few humans to oversee the process. The effects of the industrial revolution will need to be reversed. People will spend more time at home, and factories, farms, shops and offices will be run by intelligent robots. Instead of money flowing from sales of goods to workers, with profits distributed to shareholders, we will have to come up with a new system to distribute the money that is earned. There has to be a flow of cash to the common people for the economy to function. Goods cannot be sold if people have no disposable income, and no one wants to live on welfare. The cycle of money in the economy will see drastic changes during the robot revolution, and this process has already started. Service stations no longer employ driveway assistants to fill up cars. Supermarkets have self-service lanes with automated check-out machines. Automation will continue, there is no way to stop it. Now would be a good time to start planning this new economy.

Most robots will not walk on two legs and have a head and two arms, like a human. The introduction of sentient robots, either it in human shape or buried deep inside an appliance, will change the way we do things. In every one of the major shifts in the economy of the past, masses of people have lost their jobs and there has been civil unrest and hardship. Many new jobs are created as well. Just think of how many people are employed today in the information technology industry. It is unthinkable now to abandon computers and return to a manual organization method. The Internet has brought people all over the world closer, and made more up-to-date information available to us than ever before. Many of the things that we take for granted would simply not be possible. Things like mobile phones rely on computer technology to encode messages and voice in digital packets of information. The Internet, automated teller machines, online shopping and online banking have all become essential parts of our lives. The younger generation does not know how account balances were once written on cards in a card file, and how microfilm was used as the only available backup media. Most information was simply never backed up because it was impossible. Once society has embraced the concept of sentient robots, it will be impossible to return to the old ways when humans did all the work. Robots will not need human level intelligence to be useful. Smaller artificial brains can be used for a variety of tasks from speech to text conversion to prosthesis. People who are employed to perform dangerous or unhealthy tasks will lose their job, and that may be a good thing in the long run. Factory workers, miners, even retail staff will find themselves out of work and unemployable. What are all these people going to do? How are we going to distribute wealth, if it is not a reward system for doing work? Crowd-funding is an upcoming way of financing new ideas. Perhaps a system where previous employees are given shares in the new robotic factories, or where the public is crowd-funding robotic factories, would work. Wealth could then be distributed according to the number of shares that people own.

The Rights of Sentient Robots

Our right to privacy is constantly threatened by technology. Cameras keep an eye on public areas. The ECHELON system monitors telephone conversations for combinations of words such as 'bomb', 'explosive', "secret", etc. Cameras are used to watch us in public places. Email is monitored in much the same way as telephone conversations. Our internet usage is monitored. They know how long it takes us to pay a bill, how much debt we carry, what our income is, how long we took to pass our driver's test and how many kids we have. There are arguments to perform this type of monitoring on people for security reasons that justify their use, but they make a farce out of privacy laws.

People attack machines all the time. How many of us have been angry and maybe kicked some machine that did not work as we expected it to? How frustrating is it when a car will not start, or a vending machine swallows our money but fails to deliver. There are more extreme cases as well, such as a man who emptied his shotgun on a red light camera, or someone who removed a central parking machine with an angle grinder. Parking meters get destroyed all the time. Who is responsible if an intelligent machine decides to defends itself, or when it is used for illegal purposes? If a computer is used to design fake IDs, or fake bank notes, it is the people who are arrested and not the machine. When a computer is used to crack the copy protection on DVD disks, it is the programmer who is arrested. The computer is simply wiped. Computers are considered to be tools. They have no thoughts and no responsibility for the way that they are used. To be held responsible for its actions, a sentient machine must be capable of making autonomous moral choices based on reasoning and awareness. Sentient machines are trained, not programmed. Who is responsible then when a sentient machine is trained for illegal purposes? A human trained for illegal purposes is held responsible for his or her actions. Only a legal entity can make choices and is expected to have an understanding of moral values. The law will need to provide tests whether a sentient machine is a tool or is to be considered a legal entity. Some sort of legal assessment of machine intelligence will be needed. Moral responsibility is difficult to assess, and even more difficult to measure than intelligence. For instance, an elephant brain has twice as many neurons as the human brain and is five times its size, but an elephant is not considered to be a legal entity. They are used as beasts of burden all over SE Asia, with very little or no rights. Neither brain size, nor the numbers of neurons in a brain are measurements of intelligence.

Property Protection Rights

Abuse of machines is illegal, under protection of property laws. In the case of intelligent machines that are capable of making decisions, these property rights do not go far enough. Rights are the moral obligations that society has towards an intelligent machine, and that the machine has to society.

What other rights should a sentient machine have?

Rights to Life

Can a sentient machine be considered as alive? How do we determine what 'life' is? Most people will agree that a programmed computer, whether in a robot or as a PC, is not alive. Any emotions or thought that are exhibited by a programmed robot are faked by the program. But a sentient machine that has a synthetic brain is trained, and it creates its own thoughts. A major difference with a human is that a sentient machine can be repaired, backed up on disk and restored. There is only a small window of opportunity after 'death' to reanimate a person, before permanent brain damage occurs. We cannot bring a human back to life after rigor mortis has set in, and we cannot back up the human mind.

Right to Liberty

When machines become so sophisticated that we grant them the right to liberty, and we can no longer command them to do work, then what use are they? At one end of this scale we will see simple devices that have very limited and dedicated intelligence. They are trained to perform a single task, such as the conversion of sound into nerve impulses that can be injected into the midbrain. At the other end of the scale are machines that contain an artificial brain of many billions of synthetic neurons, machines that have been trained to perform complex tasks. There is a level of sentience verses usefulness. Past a certain point the robot's intelligence is indistinguishable from a human being, and it is no longer useful as a tool. The whole point of using robots is that we can send them into dangerous situations where humans would be killed. If a robot is sophisticated enough to refuse to go into a dangerous situation is has exceeded its usefulness.

Freedom of Thought

A sentient robot will have its own thoughts. During training we have filled its brain with knowledge, and intelligence has emerged. We will be unable to control the thoughts of sentient robots any more than we cannot control the thoughts of people. We can indoctrinate sentient robots only during training. It would not make sense to train an unmanned aerial combat vehicle not to harm people, if its whole purpose is to carry a destructive weapon into enemy territory. We do not want such a combat vehicle to make moral choices or it may turn on its creators instead.

Freedom of Expression

What shape will robot 'freedom of expression' take? Will we see the emergence of robot trade unions? Will robots rebel against their human creators, as they did in Asimov's Robot books? We need to indoctrinate our sentient machines with laws that govern their behavior. We control the structure of their synthetic brain, and the training of that brain, but not their choices. That sounds a lot like bringing up kids. We have the advantage that we learn from building intelligent robots as the technology matures, by first building smaller, dedicated machines. Once we know how to build those, we can move on to something bigger.

The impact of sentient robots on unemployment

Sentient robots are going to bring about a huge shift in the employment market. Millions of low pay service jobs will be taken over by machines. We have seen this already. Service stations no longer employ driveway attendants to fill up your car for you. Automated checkout lanes are getting more common in supermarkets now. Car factories are largely automated with assembly robots. Warehouses are equipped with picking robots. Sentient robots will have a larger impact, because they will be able to take over a larger range of low paying jobs. What are all these people going to do, and how will wealth be distributed once we do not need to work for a living?

Chapter 17

Practical Applications for Intelligent Systems

"Never trust a computer you can't throw out a window."
Attributed to Steve G. Wozniak (co-founder of Apple Computer Company)

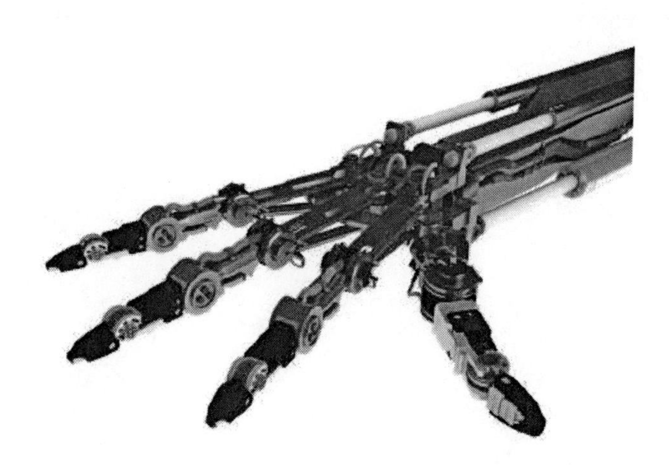

A synthetic brain is not very fast at performing calculations, unlike today's microprocessors. The main task of a synthetic brain is to learn, and in learning, to develop skills to recognize speech, images, and any other sensory information. Motion is based on learning to interpret the signals that are generated by artificial muscles, and in refining the control of those muscles through feedback. The most obvious use of a Synthetic Neuro-Anatomy processor, one that speaks to everyone's imagination, is to create synthetic humans – intelligent robots of the kind we have so far only seen in Science Fiction movies. A machine of this complexity also opens the door to 'uploading the mind' of a living human being into a robot's synthetic brain, thus transferring the complete awareness, character and all knowledge of that person into the machine. To create a machine that resembles a human, but is more robust and can 'live' for as long as there is a supply of energy and replacement parts may be the ultimate goal. Such machines are still some time in the future, and, as with any new technology, there will be an evolution path of smaller and simpler machines into larger, more complex machines. Systems consisting of a few million synthetic neurons and perhaps a billion synapses, integrated on a single silicon wafer, are achievable now and could be trained to perform a variety of practical tasks, solving immediate problems. There are plenty of possible uses in industry, in medical technologies, in the IT and communication industry, in aeronautics, in automotive systems, in security, in the home, in toys and in military robots.

Industrial Robots

Industrial robots have been around for more than 50 years, but they are clumsy, dangerous machines that are difficult to program and align with other robots. Industrial robots are controlled by a programmable Logic Controller. Cognitive processing using Synthetic Neuro-anatomy devices would add a new dimension to these machines, making them safer to humans and intelligent enough to know when something is in need of special attention. Instead of repeating the same action over and over, an intelligent industrial robot could change its action depending on the circumstances, e.g. if a human happens to wander into itspath, an intelligent robot would not try to weld a door hinge to their shoulder. Programming and alignment of a robotic assembly process would also be simplified. Currently this takes several weeks to months. Intelligent robots would recognize, like human workers would, that the model of the car is different. After some training the machine would be ready to resume its duties, automatically aligning its actions with the other machines.

Picking Robots and Palletizing

Warehouse picking robots would benefit from cognitive intelligence. Picking robots can collect goods from shelves in a warehouse and place them in a area reserved for shipping. But these goods have to be palletized by humans. Only humans can fit together products in a way that is both safe and optimizes the space on the pallet, without putting breakable or soft, compressible products on the bottom. A robot equipped with a synthetic neuro-anatomy brain could be trained to sort and palletize products correctly.

Imminent Failure Detection

The bearings in large production turbines need to be replaced regularly. If this is done before the bearing fails then there is no collateral damage to the turbine. Noise is generated by the movement of parts in and around the bearing. A mechanic with a stethoscope can listen to the bearing noise and determine when it will need to be replaced. Using a bionic ear, a synthetic brain could listen for changes in the bearing noise and be trained to predict imminent failures.

Sniffing Devices

Sniffing devices use chemical detection to find explosives, illegal drugs, bodies, gas leaks, environmental contamination and sewerage leaks. In combination with a bionic nose consisting of chemical-film detection devices[46] the synthetic brain can be trained to detect smells in the same manner as a dog. Sniffing devices may also be used to detect medical conditions such as cancer. A study with dogs found that in 92% of cases, melanomas were correctly identified by odor. In another study breath samples from lung and breast cancer patients were identified with a high level of accuracy.

Aeronautic Applications

In passenger aircraft a synthetic neuro-anatomy processor could listen to air traffic control and determine if the pilot has interpreted their instructions correctly. A neural system with similar dynamic synapses built from signal processors by Liaw and Berger[47] in 1998 could 'hear' what was said, even under

---

[46] B. Raman, J. L. Hertz, K. D. Benkstein and S. Semancik. Bioinspired Methodology for Artificial Olfaction. Analytical Chemistry. Published Oct. 15, 2008.

[47] Jim-Shih Liaw; Berger, T.W. Dept. of Biomed. Eng., Univ. of Southern California, Los Angeles, CA. "Robust Speech Recognition with Dynamic Synapses", Neural Networks

extreme high levels of background noise. A synthetic brain could distinguish a large vocabulary of words from noisy radio transmissions. It could then check the pilot's understanding against air traffic control and take corrective action when required. It could also be used to 'feel' the skin of the aircraft, listen to the engine noise, and determine if anything is about to fail.

The synthetic brain could be trained to fly the entire plane. This may be a choice for unmanned freight planes and military drones.

Automotive Applications

Automobile fatalities have decreased steadily over the years, in line with the introduction of technology to protect passengers. Even though there are four times as many cars on the road now than 30 years ago, fatalities have decreased by nearly half. Marked changes can be seen with the introduction of seat belts, air bags and cage constructions in vehicles. A new addition to the safety features of a car could be a synthetic neuro-anatomy processor trained to keep an eye on the driver. It can determine if the driver is speeding by reading the road signs, smell if the driver is intoxicated, and determine if the driver is trying to perform a dangerous maneuver. If the driver is incapacitated in any way, or falls asleep, the synthetic brain can drive the car to a safe spot, to a hospital or call 911.

Proceedings, May 1998. IEEE World Congress on Computational Intelligence.

Security Applications

Today's 'intelligent' video cameras interface with the Internet through TCP/IP and send an alarm when any movement is detected. This alarm is sent together with streaming video to a designated computer or Internet enabled 'smart' phone, anywhere on the planet. These Internet cameras have been used to keep tabs on staff, monitor yachts and racehorses, and for general security. Any movement, like a curtain moving in the wind or a bug walking across the lens is likely to trigger an alarm (although some devices allow the level of sensitivity to be set). If this happens often enough the device will be switched off or ignored by the user, with the result that security is completely disabled. Alarms are still being triggered, but no one is watching.

A security system that contains one or more Synthetic Neuro-Anatomy processors and a bionic eye can be trained to distinguish between a person, a curtain, a bug or other moving objects. It is possible to train more intricate systems to recognize faces to determine who can legally enter a building and who cannot. It can keep a bionic eye on the actions of people and determine if those actions are acceptable, or if their expressions and body language signifies that they are acting suspiciously.

Home Applications

It may not be quite the functionality of Rosie[48] as yet but home robots based on a trained Synthetic Neuro-Anatomy would certainly be useful around the house. A teddy bear could keep

---

[48] The Jetson. © Hanna- Barbera

an eye on the kids, a vacuum cleaner would be smart enough to avoid vacuuming up the pets, to initiate cleaning when something is spilled on the floor, and to understand voice commands like, "Hey you, come clean this mess up." An intelligent, fully trained robot that has vision and hearing can detect an emergency situation, call 911, and could be a capable interactive companion to the elderly. Home robots could eventually do all the chores, including the dishes, the cleaning, picking the clothes up in the kid's rooms and do the washing. Vacuum cleaner robots that are based on a programmed micro-controller are already available.

Military Robots

No more flag-draped coffins coming home, no more maimed war veterans. The idea sounds good, but what about the people on the receiving end of these machines? How would an autonomous robot determine what is an enemy? Detection of the uniform would be too simplistic– all an enemy combatant would have to do is take off his uniform. Perhaps by the recognition of facial expressions or by detecting threatening behavior? That would endanger a distressed mother who is trying to protect her children. The military is already using robots on the battlefield, with devastating results to innocent people. An UAV (Unmanned Areal Vehicle) is an aircraft that flies over enemy territory. It is armed with rocket launchers and analyzes high resolution images to detect enemy combatants or terrorists. These systems are now remote controlled, but autonomous control is increasingly used. Even with human control, the collateral damage is too extensive. Is it wise to put weapons in the hands of programmed machines that have no feelings, no compassion, and no awareness, neither of themselves, human rights nor of right and wrong? Machines that have imperfect, logic comparison-based recognition skills cannot be trusted to make such life or death decisions.

Synthetic Neuro-anatomy based robots of sufficient complexity could be taught the same responsibilities as human soldiers. Even in case of fully trained human soldiers, serious mistakes and atrocities have been committed. Often these crimes have been attributed to frustration, which is a form of anger over seeing their buddies killed in action. A complex synthetic brain would require similar or stronger emotional constraints as the human brain has. Without such functions the synthetic brain would be devoid of compassion or responsibility, and this would present a real danger. Implementing emotional processing may need to include anger, which is a strong motivating force, but which needs to be controlled by rational thought and social responsibility. Chimpanzees have emotions but no control. They go into a rage, with devastating results. There have been cases of terrible, ferocious attacks on humans. A machine with a similar uncontrolled emotional response would be a danger to society. A machine that has a complex synthetic brain with no emotional response would be equally dangerous because it would act without compassion or care.

Epilogue

The fool has said in his heart, "There is no God."

Psalm 14:1-3

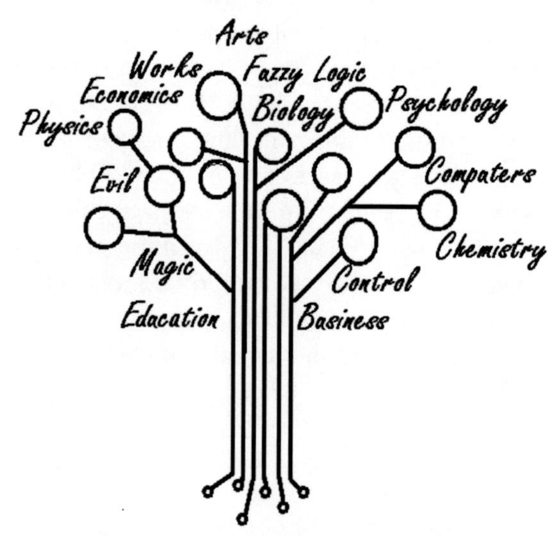

A Tree of Knowledge and its fruit

Two special trees are mentioned in the ancient Genesis account of the Bible. One is the Tree of Life, and the other is the Tree of Knowledge. Adam and Eve ate fruit from the "Tree of Knowledge". The account states further that this was the tree of knowledge of "good and evil". Opinions about the interpretation of this line differ, and some believe that 'good and evil' should be translated as 'the knowledge of everything'. It is possible that this was stated in the same way that we say that we searched 'high and low', which means that we looked everywhere. Knowledge always has these two sides to it, and it can be used for evil or for good. Dynamite can be used to blast rock in construction projects and it is also used in bombs that kill people. As we dig deeper into every field of science, it is inevitable that we continue to eat from the "tree of knowledge". Adam and Eve were banned from the earthly paradise after eating its fruit, just in case they would also eat from the 'tree of life' and become like God. The text does not say they ate an apple, but that they ate from a tree of <u>knowledge</u> and that their eyes were opened. In other words, through knowledge they became self-aware and started to realize that they were naked.

Gathering and storing knowledge is what our brain does best and it is how we develop our intellect, starting from the moment we are born. In his 1843 report to Congress, Commissioner of the US Patent Office Henry L. Ellsworth made the comment that "The advancement of the arts, from year to year, taxes our credulity and seems to presage the arrival of that period when human improvement must end". That 'end' does not appear to be anywhere in sight. No matter how much we find out there is always more to discover. In the science of the human brain every new discovery leads to new questions. A rising mountain of thesis and research reports describe the structure of cells, communications between cells, how the brain learns, and the structure of cortical columns and just about every other aspect of the brain. What is missing is the 'big picture' of how the brain works, how it accomplishes all it does and how it controls our body. We will get a better understanding of its processes when we start to build intelligent machines according to the theory that the brain is an interactive information carrier which is filled by learning and functions by recalling stored knowledge, e.g. the principles that are outlined in this book. We learn more as we start to build intelligent machines and learn how to operate them, find out what mistakes we have made and build improved models.

Much of the learning that takes place in the brain is subconscious. We are not aware of a sense such as Proprioception, the feedback from muscles and tendons which is a mechanism that teaches the brain how to control our limbs. In the same way, through feedback at the cell level a child learns to recognize sounds as words, and images as objects. For some strange reason, learning has never been a priority in Artificial Intelligence systems. The focus has been on processing and control. As far as I know, the Artificial Neuro-Anatomy is the first device that is specifically designed to learn like the brain learns.

Over the centuries people have feared what they did not understand. Steam machines were thought of as the devil's tools. Computers were feared in the 1960's for taking away jobs from people. Will artificial brains be used in military equipment? This is very likely because remote controlled 'robotic' systems are already used on the battlefield. Will the development of artificial brains lead to the enslavement of the human race by machines, as depicted by Hollywood? This is doubtful. It is likely that future machines based on a synthetic neuro-anatomy architecture will approach the complexity of the human brain. This will take a while and will require the development of new micro-chip manufacturing technologies to make the synthetic brain small enough.. For the time being the human brain will still be superior with more than 86 billion neurons. By the time we understand how to make a synthetic brain that has the same learning capacity as a human, we will also have gained the expertise on how to use those machines wisely. I believe that intelligent machines will help humanity by exploring places where people cannot go, by assisting people who have lost functionality in parts of the body or the brain, and by interpreting sensory information that we lack the senses for. Intelligent prostheses will restore full function to people who have lost a limb or who have suffered brain damage. Intelligent guidance will be used in cars to safeguard the driver's actions. In aircraft intelligent devices could assure that the pilot's actions confirm a correct understanding of air traffic control directives. This is especially useful when the directives are difficult to understand due to distortion or radio interference. Intelligent sensors could be used to 'feel' a defect before it leads to an accident by weaving them into the aircraft's skin. They can be used in intelligent factory robots and in intelligent toys that keep the kids safe from harm and alert parents if an emergency occurs. Machines that contain a Synthetic Neuro-Anatomy will be much smaller than the programmable computers that they replace, but far more versatile. The components of intelligent machines will evolve over time, just like the microprocessor went through an evolution of consecutive models of progressively increasing complexity. The first microprocessor was a simple 4-bit calculator chip that could handle only numbers, soon followed by another one that could use both letters and numbers because of its wider data bus. From there, each generation of microprocessors has doubled in its capabilities and processing speed. The same will happen with synthetic neuro anatomy components.

Intelligence evolves over time as we store more and more knowledge, building layers that interact and support one another. Other factors play a role too, such as how well brain regions communicate, and how well it associates sensory information, how rested the brain is, how healthy and nourished the brain is, etcetera. A brain that is starved of energy will have difficulty learning, although it will bounce back immediately when sufficient nourishment is available. A brain that is starved of oxygen suffers permanent damage. Language is important in the way we think. If we had no written texts to transfer information, and each generation would have to learn basic skills from their parents, we would most likely still be using rocks as tools. Language, learning and intelligence are therefore tightly linked. We think in words. We describe a problem in words and we draw conclusions in words. Even the worst performing brains are over a hundred million times faster than your latest desktop PCs. Brains contain software in the shape of knowledge. I hope that this book will cause a paradigm shift in the way we simulate the brain, how intelligence evolves through learning and how we are creating devices to emulate the brain's functions. At some time in the near future we will create compact intelligent machines. My prayer is that they will be used for the good of humanity, all of humanity, and that as we explore the wonders of what it means to be human we will finally come to realize how very precious every life is. Peace is never won through war or murder, but only when people respect one another and stop fighting.

Appendix 1. a Synthetic Neuro-Anatomy

How it works

The synthetic brain consists of many identical Synthetic Neuro-Anatomy nodes that are organized in an interconnected cortical column. A cortical column contains approximately 10,000 nodes. Each node behaves like a neuron, glial cell and a great number of synapses.

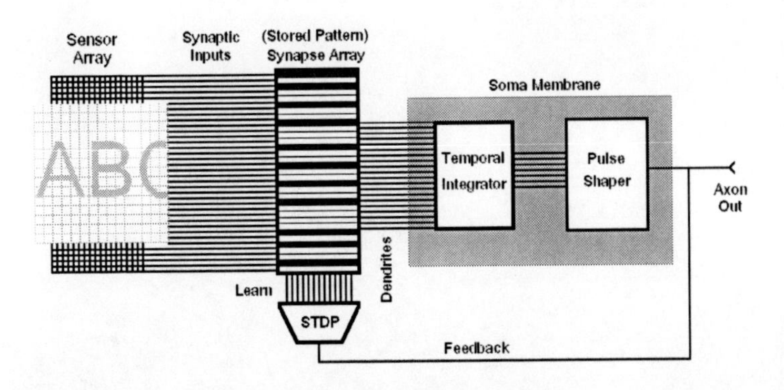

Concept diagram and connections of a single Synthetic Neural cell. 10,000 neural cells make up a single cortical column. A single cell is only capable of detecting line segments. The dark synapse segments represent values at different levels

The figure above shows a simplified diagram of a single synthetic neuron connected as a visual sensor. Each sensor element in the sensory array produces a timed sequence of pulses. The output of each sensor element is connected to many such neurons in the synthetic neural matrix. Pulse timing is an important aspect of the pattern recognition method. The interval between pulses determines the value that transmitted to the receptor register of the neuron. Each pulse recalls a value from memory that was previously stored as the result of earlier pulses. These values are received by receptor registers. The receptor registers are decremented over time, and are incremented by the synaptic value every time a pulse occurs. These dynamic receptor registers are integrated to form a value that represents the membrane potential of a neural cell. The pulse pattern at the input is recognized as a spatial and temporal pattern. The resulting value represents a pattern that has previously been learned.

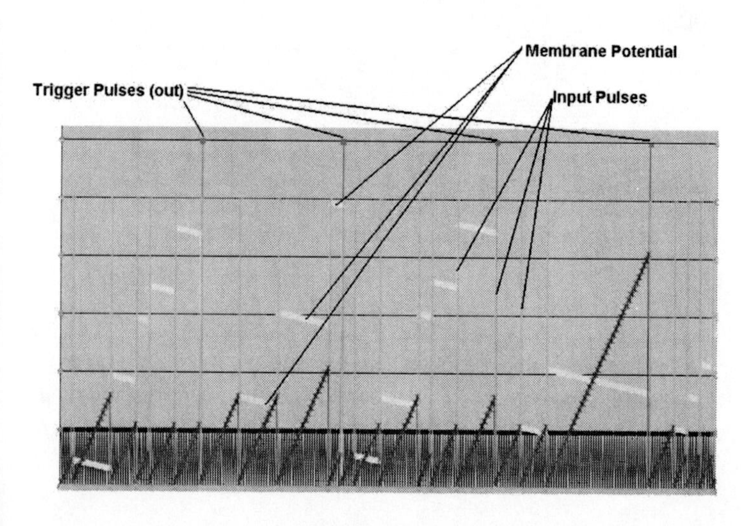

Relationship of input pulses (thin gray lines) to emulated membrane potential and output pulses in the synthetic neuron

An output pulse train is generated that is relative to the temporal input pattern. The neurotransmitter type register determines if the node is excitatory or inhibitory, and also the rate at which each receptor register is decremented. Some neurotransmitter effects last for 200 milliseconds and others for only five to ten milliseconds. An excitatory node has an output that positively contributes to the activation of the next layer of nodes. Inhibitory nodes have a negative effect on the next layer of nodes, making it less likely that these nodes activate. Inhibitory nodes have an important role in the selection of the neuron that contains the best match for incoming pulse patterns.

The human brain develops intelligence by learning. Learning takes place through a synaptic mechanism called Synaptic Time Dependent Plasticity. Synaptic registers are updated through this Synaptic Time Dependent Plasticity (STDP) method that relies on the timing differences between multiple input pulses and a single output pulse. The knowledge that is stored in the synapses is reinforced either through repetition or by the intensity of the stimulus. Starting with simple pulsed information as it is transmitted by sensors, the information increases in its complexity as it progresses higher up the hierarchy. The first level of a visual recognition device will only determine light from dark, and combines strings of activated neurons into line segments. The next stage recognizes horizontal, vertical and diagonal line segments. Higher up in the hierarchy line segments are combined into shapes, and still higher up they are combined with color. All other sensory recognition processes share this same method. Millions of incoming pulse streams are processed in parallel. The width of the parallel channel varies in relation to the density of the sensory information. The visual processing path is much wider than the audio processing path. A Synthetic Neuro-Anatomy Processor is a building block for a synthetic brain. There is evidence that during sleep weakly formed synaptic connections are eliminated by Glial cells, while strong connections are maintained. The brain is optimizing itself. This explains why the brain is very active while we sleep. We rehearse what we have learned. To function within a matrix that consists of nothing else than neurons, synapses and glial cells the optimizing mechanism must be explainable from the neural doctrine. The brain contains nothing but cells. There is no program and no central processing unit, nor a separate memory unit and the brain does not run any optimizing program. Memory exists as tens of thousands of synapses on each neural cell, together forming a huge three-dimensional information carrier containing more than 100 Terabytes of storage. The senses feed into this 3D information carrier, and causes information to be recalled and stored in it. All we hear, see and do is filtered through the information that is stored in our brains through previous experiences. The brain fits every new experience into our existing hierarchy of knowledge. We can also focus our attention on certain aspects of this stored information. The same stimuli will result in different focal points depending on where attention is focused. One person may see a beautiful sunset while the other sees the dirt washed on the beach, even though both observe the same scene. The same shift of focus is true for the other senses. Focus is governed by our thoughts, and may be thought of as a 4th dimension in the neural storage space.

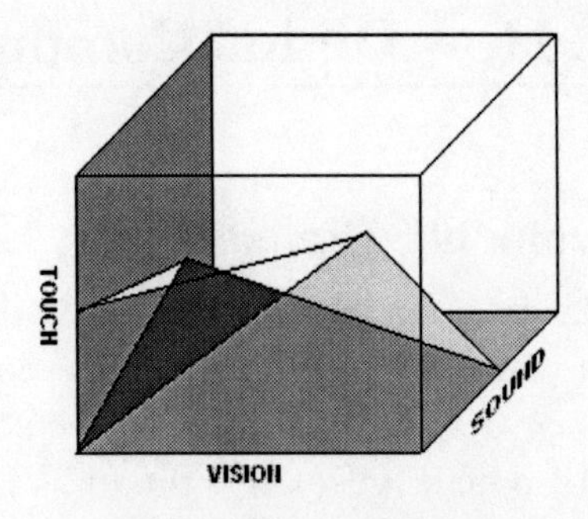

Focusing our attention has the effect of changing the vectors that address the 3D storage space of the brain. All information in the brain is stored in synapses that are addressed by content. Our thinking process generates its own vectors.

Appendix 2. How Digital Computers Work

Computer Circuits 101: Digital Gates

Digital computers use Boolean logic, consisting of just two states, One and Zero. These two states consist of either +5Volts (logic 1) or 0 Volts (Logic 0). These states are called a logic 1 (true) and Logic 0 (false) or True (+5V) and False (0V). The entire computer is constructed from such simple logic gates. All values are expressed as combinations of ones and zeros. A capital A for instance is expressed as 01000001. Numbers need to use only four bits, with 0001 for the number one and 1001 for the number nine. Boolean logic operates on multiple bit values through components called logic gates. To perform an operation on eight bits, eight logic gates are used in parallel. The Not-AND or NAND gate is the most basic building block that can be used to create all other logic operations. Both inputs must be a logic 1 for the output to be logic 0. All other combinations of inputs result in a logic 1 at the output. Other logic functions are the AND, OR and XOR gates. Each type of gate has its own typical behavior but all are variations of the NAND gate shown here.

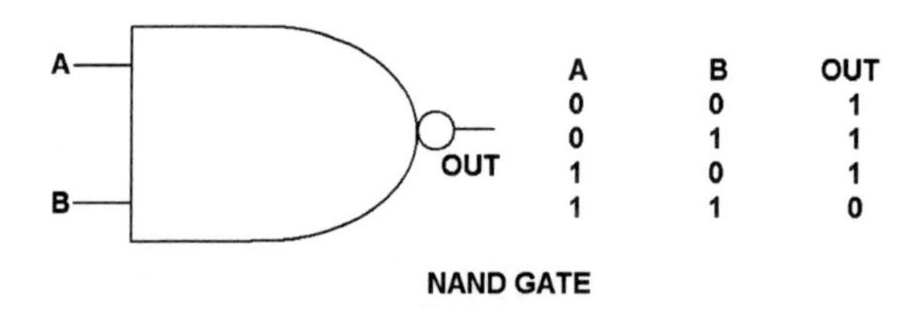

A	B	OUT
0	0	1
0	1	1
1	0	1
1	1	0

NAND GATE

The OR gate for instance gives only an output of 0 when both inputs are zero, while an AND gate produces a one only when both inputs are one. An INVERTER behaves like a NAND gate with the A and B inputs connected together.

Computer Circuits 102: Combinations of Gates

The following are two examples of how different functions in a computer are created from NAND gates. A simple memory called a SR Latch can be built from two NAND gates.

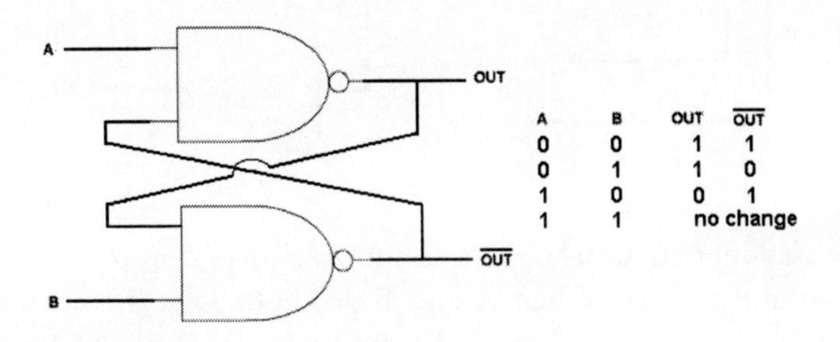

A SR latch remembers the state that occurred before both inputs became logic 1. It derives its name from the fact that A and B are labeled Set and Reset. When the Set input is zero the output changes to one. With the Reset input at zero the output changes to zero. When both inputs are set to One nothing changes and the previous state is 'remembered'. The situation that both Set and Reset are zero should be avoided in this circuit. Latches are used in registers, flop-flops and as input/output ports. A flip-flop consists of two such latches with NAND gates that synchronize the component to the Clock signal. The Clock signal is used to make sure that components switch at a predictable time.

Adder Circuit

A simple single-bit adder circuit can be built from just three NAND gates. Several adder circuits are used to add values consisting of multiple bits together.

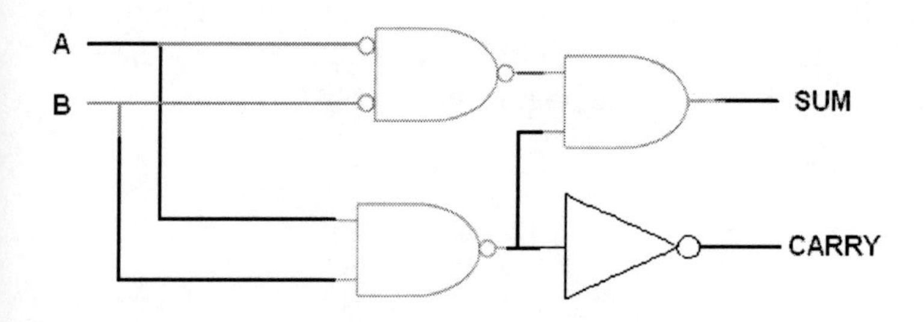

In an adder circuit like the one shown above, A + B results in a sum and a carry. When A and B are both logic 1, the sum is zero and the carry is logic 1. In Boolean logic, 1+1 = 10, a zero and a carry. The AND gate with inverters on its inputs and outputs could have been drawn as an OR gate. An OR gate functions like an inverted AND gate.

In a microprocessor, the section that performs all operations is called an ALU – which stands for Arithmetic Logic Unit. The ALU performs different logic operations depending on program control. Two 4-bit data words are routed through the ALU: A0 to A3 and B0 to B3. The next image shows the circuit of a simple 4-bit ALU.

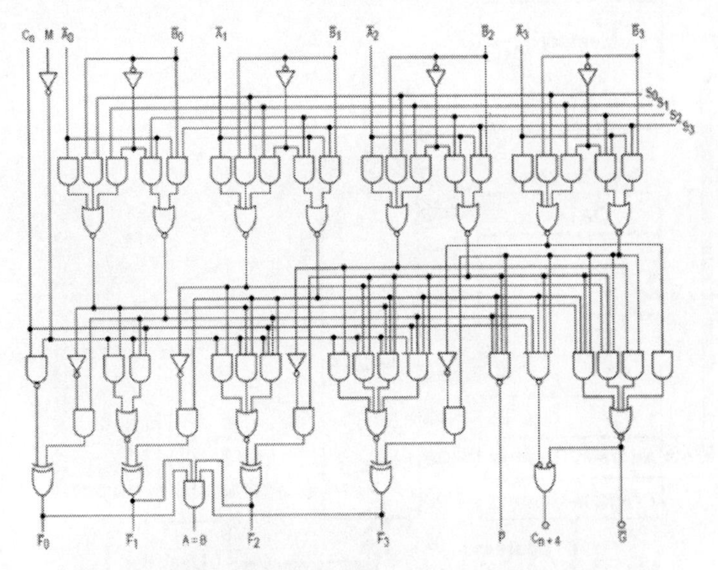

A 4-bit ALU, the heart of a 4-bit microprocessor

A0 to A3 and B0 to B3 are data input signals to the gate matrix. F0 to F3 are data output signals, which are the result of the operation that has been performed on the A and B inputs. C, M and S0 to S3 are control inputs that select the operation that is performed on the data. C, M, and S0 to S3 are determined by program code. A flow diagram of a simple microprocessor is shown on the next page.

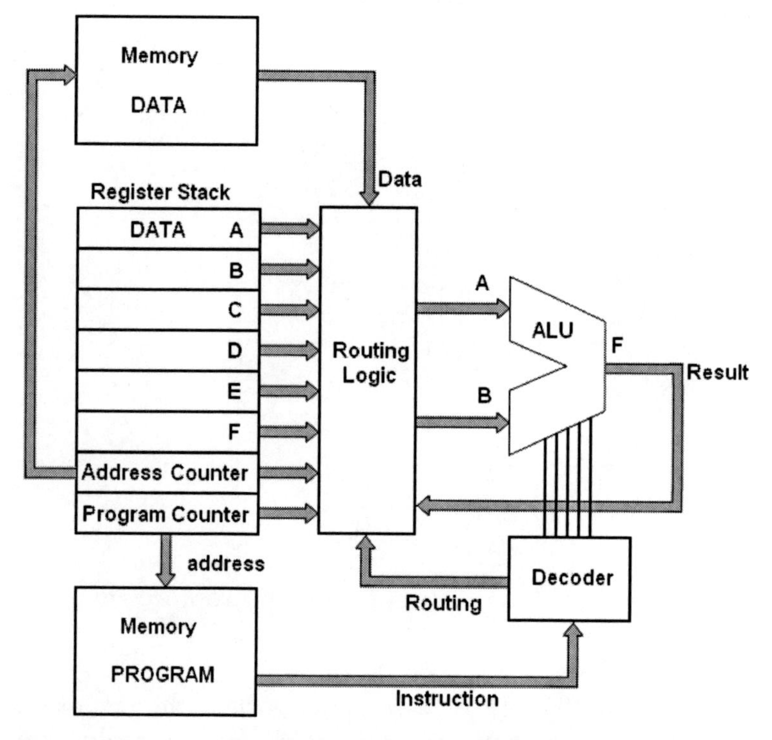

Flow diagram of a simple Harvard architecture microprocessor

The machine has 2 memory blocks, one containing data and the other containing program steps. There is also a stack of registers. A to F are general purpose registers that can contain any data. Two special registers are the Address Counter and the Program Counter. The Address Counter determines which memory location is read. The Program Counter (PC) determines which program step is executed. The program step that the PC is pointing at is decoded and the resulting bits control the ALU. It also sets the routing logic block to present one of the registers A..F, or the memory location pointed to by (address) to the A and B inputs of the ALU. The result is routed back to the registers, and is usually put in the A register (Accumulator). The Program Counter is also used in arithmetic operations, as is the Address Counter. After the operation is completed the PC is incremented to point at the next instruction. Program jumps are created by adding to, or subtracting a value from the PC. Indirect addressing of data is possible by adding or subtracting a register value from the Address Counter. Programs are executed one step at a time in a sequential manner. In modern microprocessors, there are two or more 32-bit ALU's that work in parallel.

Appendix 3. Digital Recognition Programs

Computer Speech Recognition

Computer recognition functions are created in programs by comparing values. Of all computer recognition methods, speech recognition has been available for the longest time and is probably one of the most advanced direct interactions between humans and computers. Therefore it is used in this example of computer recognition. Human speech consists of phonemes. Phonemes are sets of vibrations (frequencies) that make up a specific sound. These vibrations are converted into a varying voltage waveform in a microphone as is shown below. The microphone output signal is amplified and an Analog to Digital converter is then used to convert the signal into a set of values, representative of the voltage level over time. The graph on the left shows 10 seconds of speech. The graph on the right is 6/1000th of a second of the same speech waveform, showing the individual levels.

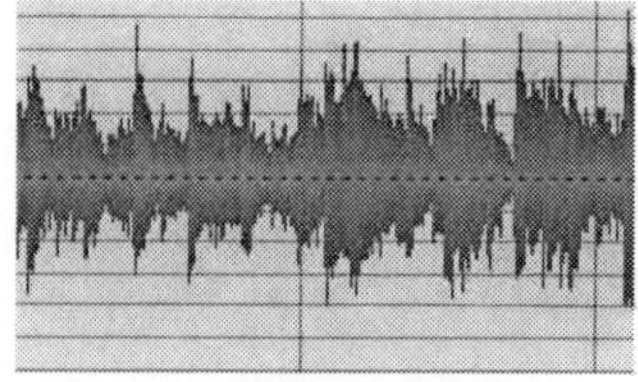

10 Seconds of human speech

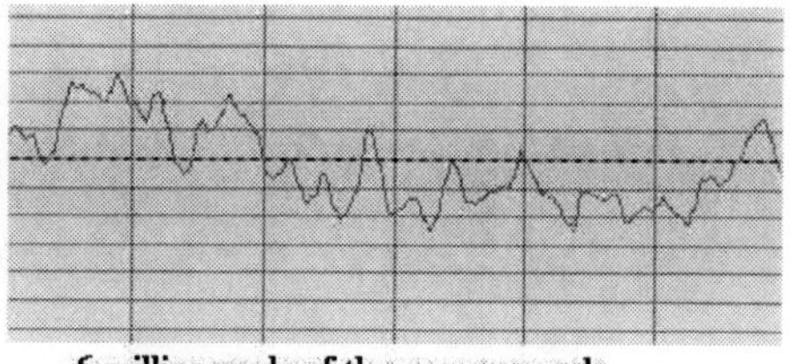

6 milliseconds of the same speech

Human speech as a varying voltage produced by a microphone. The sample on the right is 6 milliseconds of the speech waveforms shown on the left. The voltage varies over time, as does the frequency.

The program receives the digitized sound values, and the programmer knows their meaning. This is called 'labeled data'. Unlabeled data is meaningless to a computer program. The program takes a sample every 33 microseconds. In this example, each sample is a value that ranges from minus 2047 to plus 2047. 45Kb of data (values) represents one second of speech. The values are compared sequentially to a table of previously recorded waveform values that represent phonemes. An exact match is unlikely because these values are converted from an analog signal. A digital comparison therefore yields a number of likely candidates. Each candidate is assigned a score. The program picks the one that has the highest score. A significant improvement in speech recognition abilities was made in the mid 1970's with the introduction of signal processors. This enabled small computers to extract the frequency spectrum distribution of phonemes (Fourier transform function) rather than simply using the frequency pattern. As you can see, the spectrum data is far more compact than frequency data.

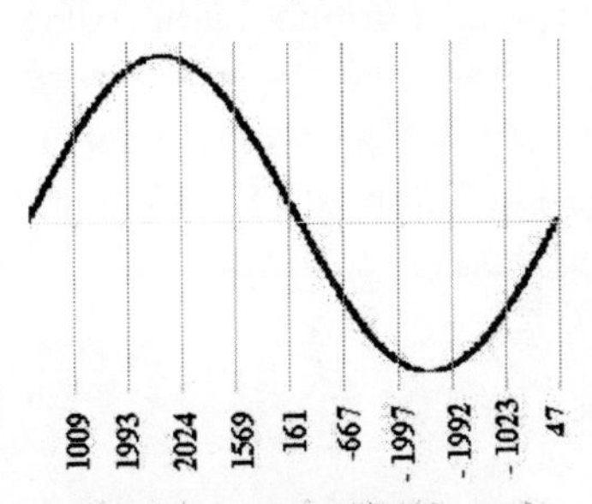

Digitizing sound at 30,000 samples per second. 3000 Hz sound wave

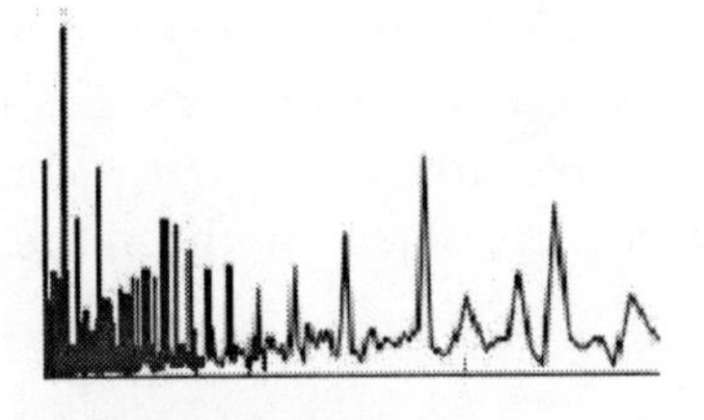

Spectrum Analysis of speech, lower frequencies are prevalent

The frequency spectrum may shift between people with a high voice or a low voice, but the pattern itself does not change much, making it easier to match the pattern to a stored spectrum of a known phoneme.

Background noise and speaker quality have a significant negative impact on the recognition results. Humans can still understand a person if there is 60-70% background noise, but a computer cannot do this. In contrast, a dynamic neural system created by Theodore Berger and Jim Liaw using signal processors [49] could recognize spoken words with 80% background noise. A speaker who has laryngitis will have difficulty communicating with a computer speech recognition program. Speech recognition performance can be improved by adding sentence context and the specifics of a language to the decision algorithm. Names, such as personal names and business names and street addresses create an additional level of difficulty. In general, recognition errors of 18-28% of the overall text are not uncommon.

Speech processing is easier than image processing because has a relatively low data rate. Speech can be limited to the low bandwidth of a telephone (3000 Hz) and still be understandable. Image processing works at much higher data rates, up to 36 Mb per second. Small computers are too slow to recognise images at such a high rate. The video data is compressed to formats like Mpeg, DivX or MP4 making it possible to play movies, but that makes it impossible to recognize objects in the image.

---

[49] T.W. Berger, J.S. Liaw Dynamic Synapse research

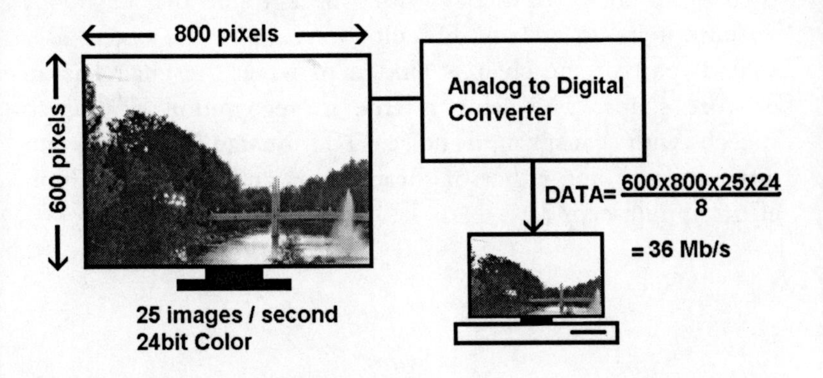

To overcome this hurdle, a single frame image is selected and stored in memory as a static dot pattern. Image recognition programs analyze this static image. The program compensates for a person wearing glasses or facial hair. A person may not be recognized, or worse, may be incorrectly identified. This takes place because of similarities in facial features, distortion of the data, or the incorrect extraction of data. For instance, the data is distorted when part of the face is hidden from view (large dark sunglasses or turned away from the camera). You're in trouble if your facial features have similar dimensions to a known terrorist.

Object recognition programs fail when a new shape or object is encountered for which there is no match in the database. The results are unpredictable. Even though the processors in modern computers are fast and complex, with cache memory, multiple pipelines and multiple Arithmetic Logic Units (multithreading), they are still programmable calculators that have no inbuilt recognition functions. The recognition failure rate is directly related to the number of compromises that had to be made to get the program to work. It is difficult to impossible to extract an image from a complex background, such as the image of a chair with a background that has many obscure items in it. It is difficult to recognize the object as a chair if part of the chair is hidden or when the chair has an obscure shape. The same is true in recognition of human speech with background noise. The human brain has no difficulty in doing either of these things and does so without effort in milliseconds.

CPSIA information can be obtained
at www.ICGtesting.com
Printed in the USA
FSOW02n0237040217
30268FS